THE LOST
MYSTERIES OF EGYPT
A JOURNEY FROM
ATLANTIS TO ALIENS

DR. LEO RASTOGI

Anahata Books
The Woodlands, Texas

1st Edition

Paperback ISBN: 978-1-7358714-5-5

www.leorastogi.com

This book is dedicated to the timeless tradition of mystery schools and lineages of great teachers who, across the ages, have curated, protected and passed on the sacred esoteric wisdom to future generations

Acknowledgments

I would like to begin by thanking my parents, Dr. Aditya Rastogi and Dr. Rekha Rastogi. Inheriting genes from a historian and a political scientist decidedly sowed the seed in my early years of exploring ancient mysteries and their influence on our current society.

I must thank all the wonderful people in my life I have been blessed with, who have supported this journey of exploration of mysteries in their uniquely varied ways, especially by never bringing up that, "*I was going off the deep end*" – Richa, Nirjhara, Javi, Charlie, Rodrigo, Ale -- thank you.

Finally, a big heartfelt gratitude to my dear friend and editorial partner, Wendi. Thank you for sacrificing so many evenings of watching sunsets in paradise, all to help me meet my deadline. Thank you, Dmitry, for making the book look beautiful and your patience with endless revisions!

Most of all, I would like to close by thanking my invisible teachers, who enabled me to have access to halls of wisdom of the mysteries, allowing me sight not only into a world beyond sight but also into the infinite cycle of time that holds mysteries of our past and future. A lifelong humble student of yours, I remain.

Author's Note

Egypt has always held a special place in my heart, captivating me from a young age. What began as an interest in its iconic pyramids and mysterious hieroglyphs quickly grew into a personal journey of discovery, both physical and intellectual. I've spent countless hours wandering the ancient sites — from the towering obelisks of Karnak to the hidden tombs of the Valley of the Kings — and each step has drawn me deeper into the enigma of Egypt's past. But it wasn't just the thrill of discovery that kept me coming back. My own background in Vedic studies and explorations into ancient civilizations have revealed something even more profound: the continuous thread that weaves human history together, from the Vedic civilization to Egyptian, and from Egyptian to Greek.

In my travels and studies, I have found echoes of shared wisdom and symbols that transcend borders and time. The rituals, the cosmic beliefs, the understanding of life, death, and the divine — they all seem to speak the same ancient language, stretching across continents and centuries. The sacred texts of the Vedas, the mysteries of Egyptian religion, and the philosophical inquiry of the Greeks are not isolated events but parts of a larger human journey. It's as though the pieces of a grand mosaic have come together in my mind, showing a continuum of thought and spirituality that links these ancient worlds. Egypt, for me, became the central point of this convergence, a place where civilizations not only met but also transformed one another.

This book is a reflection of that realization — a desire to uncover not only Egypt's lost mysteries but to see how they fit into the broader narrative of human history. Through the lens of my own experiences and research, I hope to reveal how the wisdom of one civilization flows into the next, carrying with it the essence of humanity's quest for knowledge, truth, and connection with the divine. My hope is that, as you journey with me through Egypt's ancient past, you too will begin to see these connections, to feel the continuity of human experience, and perhaps, like me, appreciate the beauty of this mosaic that tells the story of us all.

In the chapters that follow, we will explore some of Egypt's most intriguing enigmas: the architectural genius behind the Great Pyramid, the hidden meanings in the ancient temples, the sacred texts that aimed to guide souls through the afterlife, and the mysteries of the pharaohs who sought immortality. We will dive into the controversial and mysterious — the possibility of extraterrestrial influence, the legends of Atlantis, and the theories that suggest Egypt may have been part of a forgotten, advanced civilization that once spanned the globe. Alongside these mysteries, we will examine how Egyptian thought influenced Greek philosophy, connected with Vedic wisdom, and echoed in spiritual practices across ancient cultures. This journey will take us deep into the heart of Egypt's spiritual and intellectual legacy, unveiling truths that have long been hidden in plain sight.

In sharing this, I offer you not only a glimpse into Egypt's forgotten wonders but a chance to see the world as I've come to see it — a kaleidoscope of shared wisdom, glistening in the sands of time.

Contents

Contents

"To unlock the secrets of Egypt is to wander through the corridors of eternity, where shadows of gods whisper the ancient truths of the universe, and the line between myth and reality dissolves into a timeless enigma."

Dr. Leo Rastogi

1. Introduction: Unveiling the Enigma

Are you ready to embark on a journey through the sands of time, to a land where one of the most mysterious and awe-inspiring civilizations in history was born?

We are traveling to ancient Egypt, nestled in the north-eastern corner of Africa, along the life-giving waters of the majestic Nile. From this fertile cradle emerged a people profoundly intertwined with the mysteries of life, death, and the divine — a civilization whose astonishing achievements and spiritual wisdom continue to captivate and fire our imaginations to this very day.

From the majestic pyramids of Giza to the enigmatic Sphinx, the ancient Egyptians left an indelible mark on history, showcasing their mastery of engineering and devotion to the afterlife. Their intricate hieroglyphic writing system, exquisite artwork, and complex structures provide a glimpse into a civilization that not only existed but thrived along the banks of the Nile river thousands of years ago.

When we think of the ancient Egyptians, we are immediately filled with a sense of awe, wonder, and intrigue. We marvel at their architectural prowess, their deep understanding of astronomy and mathematics, and their intricate burial rites.

Everything they left behind serves to remind us of their sophisticated culture—one immersed in the visible and the mysterious, the natural and the supernatural, the magical and the marvelous. Their worldview was shaped by chaos and order, divine ordinance and human aspirations, marked by their unique religious beliefs.

With their vast pantheon of powerful gods, enigmatic rituals, epic creation myths, and mysterious initiation rites, the ancient Egyptians were a culture grounded in profound wisdom about the nature of the world around them. Their extensive knowledge of astronomy, architecture, engineering, agriculture, medicine, mathematics, and art was interwoven with their highly developed spiritual and social structures, driving the development of a civilization that was able to successfully sustain itself over millennia.

While archaeologists have been investigating the passing of this great people for centuries and do agree on many interpretations of their findings, a lot of debate continues to surround the wheres and the whys. Researchers continue today to be intrigued, and indeed, mystified by the sheer range of knowledge these ancient desert dwellers possessed about the night skies, not to mention their ability to erect huge monuments that perfectly aligned with astral bodies.

In this book, we will be exploring many of these enigmas, delving into the mysteries of a people who seemed to be illuminated by extraordinary insights and wisdom—a wisdom that appears to have been lost in the annals of time.

There are many aspects of ancient Egypt that are still shrouded in mystery and I intend to highlight some of these in the following pages. I cannot promise to solve any of the riddles but hope that this exploration will allow you to consider alternative possibilities to age-old secrets.

For instance, how did they develop such precise methods used for their monumental structures like the Great Pyramids of Giza, which reflect their advanced engineering skills? Were these architectural marvels a testament to their indepth engineering knowledge, portals to greater mysteries, or extraterrestrial blueprints? The construction methods employed by the ancient Egyptians continue to puzzle archaeologists, with questions remaining about how they transported and lifted enormous stone blocks, achieved precise alignments, and maintained structural integrity over time. While various theories have been proposed, the exact techniques and logistics involved remain a subject of hot debate and investigation.

Also, why did they develop such complex beliefs and intricate rituals surrounding death and the afterlife, as evidenced by the elaborate tombs and funerary practices discovered in the region? Many aspects of their religious beliefs, such as the nature of the afterlife, the role of specific gods, and the symbolism of ritual objects, remain enigmatic. Even though Interpretations of religious texts and symbols are ongoing, new discoveries continue to make us question what we thought we knew of these ancient people.

How did they acquire such exact knowledge of the night skies, which they embedded into their calendar system and temple alignments, hinting at a profound connection between the cosmos and earthly affairs?

Why is it claimed that the majestic Sphinx guards under its feet the 'Hall of Records', in which is preserved the entire knowledge and wisdom of a lost civilization? If it is true, why has it never been found?

And what of the notion that the location of Plato's tale about the lost city of Atlantis can be traced back to an area once flooded by the great River Nile? What evidence has been uncovered that can shed light on this legendary city's location? I will devote a chapter related to this topic, which remains one of the most controversial flood myths in history, and hopefully offer some additional insights on the matter.

Another theme we will be exploring is the quest for lost knowledge hidden within the sacred texts and hieroglyphs of the ancient Egyptians. What revelations about their relationship with the gods are waiting to be gleaned from such cryptic texts and stories? Perhaps they hold the key to explaining the immense influence that divinities like Ra, Osiris, and Isis held over their human devotees. While the ancient Egyptian hieroglyphic writing system has been deciphered to a large extent, there are still gaps in our understanding, particularly regarding certain symbolic and religious texts. The meanings of some hieroglyphic signs and the nuances of ancient Egyptian grammar and syntax continue to perplex scholars today.

You may also be intrigued by popular ideas surrounding Egypt's extraterrestrial connections and Ancient Astronauts theories. Were the Gods actually ancient aliens and, if so, what evidence do we have of such extraterrestrial encounters? This notion may have come under a fire of criticism in recent times, yet there are still many scholars out there who are intrigued by the possibility and we will be exploring the different opinions on the subject.

Ancient Egyptian culture certainly included esoteric and mystical practices, such as alchemy, magic, and spiritual rituals. While some texts and artifacts offer glimpses into these practices, the full extent of their knowledge and its significance is still a matter of speculation and intrigue. Being a culture that was so immersed in mysticism and occultism, it's no wonder that some of its ideas filtered through into the realm of new-age movements, fuelling a desire to uncover its long-lost wisdom. We will be taking a look at the impact of Egyptian culture in our world today as we continue to be enthralled by its hidden pearls of wisdom.

These mysteries, and many more, serve as a powerful reminder of the depth of ancient Egyptian wisdom and the enduring allure of their civilization. From the towering pyramids to the cryptic hieroglyphs, Egypt remains shrouded in layers of esoteric knowledge, much of which still eludes modern understanding. The ancient Egyptians were not just masters of the material world — they were also deeply attuned to the spiritual and metaphysical realms, as seen in their intricate religious beliefs, obsession with the afterlife, and the elaborate rites that point toward their intimate knowledge of the unseen forces of the universe. Their civilization seems to have been a bridge between the physical and the mystical.

Ancient Egypt's ties to the mysteries and occult traditions cannot be overlooked. For centuries, Egypt has been revered as the cradle of hidden wisdom, a source of inspiration for countless mystical orders, secret societies, and occultists throughout history. The ancient texts, from the *Book of the Dead* to the Hermetic writings, are filled with allegories of transformation, resurrection, and cosmic laws that invite deeper reflection. What role did these teachings play in the evolution of their spiritual understanding? Could it be that they accessed hidden dimensions of knowledge, passed down through priesthoods or channeled from beings beyond our comprehension?

Throughout the chapters that follow, I will be posing questions that may forever remain unanswered, but they will undoubtedly ignite conversations around the enduring enigmas of ancient Egypt. As we delve into these mysteries, we must consider not just the historical and archaeological evidence, but also the profound spiritual and mystical currents that continue to ripple through time, calling us to explore the hidden truths that still occupy our imagination. Egypt, it seems, is not only a land of the past — it may hold the keys to unlocking deeper, universal mysteries that transcend time itself.

The ancient Egyptians have been a source of wonderment down through the ages, as well as the focal point of rigorous scientific and archeological research. Despite this, such questions linger in the air like mysterious mist—never fully clearing to allow us into their world. With each new discovery or eureka moment, the true nature of this mysterious civilization continues to slip through our fingers, like shifting desert dunes; elusive grains of sand we simply cannot grasp.

This book is a journey into the mysterious and unknown, where we will tread upon sacred paths once walked by the ancients, unravel timeless myths that continue to captivate the human soul, and examine historical evidence alongside alternative theories that challenge conventional understanding. As we venture deeper, perhaps we may uncover glimpses of the wondrous world of the ancient Egyptians and partake in the wisdom they so carefully guarded — wisdom that whispers to us across millennia, urging us to look beyond the surface and into the deeper currents of knowledge that shaped their extraordinary civilization.

Who were the ancient Egyptians, truly?

How did they attain such unmatched mastery over architecture, science, and spirituality?

What hidden secrets does their enduring legacy still hold, waiting to reveal profound truths about humanity's place in the universe?

It's a fascinating journey that I can't wait to share with you!

A Vedic student in Egypt

In the quiet of desert sands, I sit,

As ancient winds whisper truths half-forgotten,

Under the vast, endless sky of Egypt,

I feel the hum of ages, like echoes from the Vedas,

A rhythm older than words, older than thought.

The pyramids rise like prayers in stone,

Each block a mantra carved in silence,

Much like the hymns sung by sages long ago,

On the banks of Sarasvati's forgotten flow,

Where the sacred fire mirrored the stars above.

In these sands, I hear the same murmur,

That speaks of gods, of light, and of realms unseen.

The wisdom of Ra and the breath of Brahman,

Merge into one — timeless, infinite, untouchable.

The lotus blooms in the Nile, as it does in the Ganges,
Each petal a layer of truth peeled back,
Each river a thread in the web of creation.
Whether under Egyptian stars or Vedic chants,
The soul seeks the same — a journey homeward.

In this mystique, I see no borders,
No land too far, no sky too wide.
For Egypt's gods and the truths of the Vedas,
All whisper the same eternal secret:
We are but dust, seeking the light from which we came.

And when we depart, death is not an end,
But a passage — a return to the source.
Osiris guides the soul through the shadows,
As the soul, in the Vedas, flows through cycles of rebirth,
Both paths lead us back, only to begin again —
Reborn like the sun after a long night,
Like the lotus rising from mud to bloom once more.

And in whispers of Atlantis, submerged in time,
The same mystery lingers — a world forgotten,
A wisdom lost beneath the waves, like souls,
That sink and rise, finding form in each new age.
Atlantis, Egypt, and the truths of the Vedas are but mirrors,
Reflecting the same truth:

We are eternal, spiraling through birth and death,
Searching, always, for the light we once knew.

As I sit beneath the stars, beneath the ancient gaze of stone,
I feel small, humbled, moved.
In the vastness of it all — the timeless truths,
I am just a traveler passing through,
Touched by the mysteries, by the depth of life and death,
And grateful for the moment, for the knowing that I, too,
Am part of this great, infinite dance.

A Brief Timeline

- Predynastic Period - from c. 6000 BCE - 3100 BCE
- Early Dynastic Period - c. 3100 BCE - 2686 BCE
- Old Kingdom - c. 2686 BCE - 2181 BCE
- First Intermediate Period - c. 2181 BCE - 2055 BCE
- Middle Kingdom - c. 2055 BCE - 1650 BCE
- Second Intermediate Period - c. 1650 BCE - 1550 BCE
- New Kingdom - c. 1550 BCE - 1070 BCE
- Third Intermediate Period - c. 1070 BCE - 664 BCE
- Late Period and Ptolemaic Period - c. 664 BCE - 30 BCE
- Roman and Early Christian Period - c. 30 BCE - 4th century CE
- Islamic Conquest and Influence - 641 CE onward

Overview

While you may be all too familiar with well-known Egyptian figures such as Queen Cleopatra, it's important to note that the rise of Egyptian civilization began thousands of years before she had even sat on the throne in 51 BCE, with the unification of Upper and Lower Egypt by King Narmer in around 3100 BCE. Nearly 1,500 years before Tutankhamun ruled the land, and around 600 years before the pyramids of Giza were built, the foundations of what we now refer to as ancient Egypt were being laid. By this time, Egyptian civilization was emerging, transitioning from scattered Neolithic communities into a unified and organized society that would eventually flourish into one of history's most remarkable cultures.

♦ **Predynastic Period - from c. 6000 BCE - 3100 BCE**

The emergence of early settlements along the fertile banks of the Nile. This era was marked by a significant transformation in human society, as communities transitioned from nomadic lifestyles to more settled, agrarian ones. The rich alluvial soil of the Nile Valley provided an ideal environment for the cultivation of crops, such as wheat and barley, and the domestication of animals, including cattle, sheep, and goats.

The development of agriculture allowed for surplus food production, which, in turn supported population growth and the establishment of permanent villages. These villages became the nuclei of emerging social structures and hierarchies, laying the groundwork for the sophisticated civilization that would later flourish.

Religious and spiritual practices also began to take shape during the Predynastic Period, with early Egyptians

developing rituals and beliefs centered around natural phenomena, fertility, and the afterlife. They constructed simple
shrines and worshiped at sacred sites, indicating a growing
sense of community identity and shared belief systems.

As the Predynastic Period progressed, early settlements
along the Nile became increasingly interconnected. The
gradual unification of these communities laid the foundation
for the rise of powerful chiefdoms and proto-kingdoms. By
the end of the period, significant political and social con-
solidation had occurred, setting the stage for the eventual
emergence of the Pharaonic state.

◆ Early Dynastic Period - c. 3100 BCE - 2686 BCE

Unification of Upper and Lower Egypt under the rule of
Narmer (also known as Menes), which was a pivotal moment
in ancient Egyptian history. This event marked the beginning
of a time when Egypt transitioned from being two separate
kingdoms into a single, centralized state under the rule of
a single king, or pharaoh. Although Narmer is traditionally
considered the first Pharaoh of unified Egypt, the exact
details of his accession to power remain somewhat shrouded in mystery.

What we do know is that the unified land had the Pharaoh at
its helm, who was not only a political leader but also a divine
figure, believed to possess a direct connection with the
gods. This concept of divine kingship, wherein the Pharaoh
was seen as a living god on Earth, was not something new to
the region, and it became a central tenet of Egyptian society
and governance for thousands of years to come.

Having established a centralized state, the early Egyptians
were then able to undertake large-scale construction
projects, establish trade networks, develop sophisticated

administrative systems, and create a distinct cultural identity that would endure for millennia. Narmer's reign marked the dawn of a new era in Egypt, characterized by stability, prosperity, and the flourishing of ancient Egyptian civilization.

- ◆ **Old Kingdom - c. 2686 BCE - 2181 BCE**

Construction of the monumental pyramids at Giza, as well as the development of a complex bureaucracy. The Great Pyramid of Khufu (Cheops), the Pyramid of Khafre (Chephren), and the Pyramid of Menkaure (Mycerinus) stand as enduring symbols of ancient Egyptian ingenuity, engineering prowess, and religious beliefs. Their construction alone required meticulous planning, organization, and immense labor resources, reflecting the centralized authority and the ability of the state to mobilize labor for such monumental projects.

- ◆ **First Intermediate Period - c. 2181 BCE - 2055 BCE**

Characterized by political fragmentation and instability, giving rise to regional power struggles. This period saw the decline of central authority, with local rulers, known as nomarchs, asserting control over their territories, leading to a breakdown in the unity of the Egyptian state.

- ◆ **Middle Kingdom - c. 2055 BCE - 1650 BCE**

Marked by the reunification of Egypt under the rule of Mentuhotep II. During this period, there was a resurgence of central authority, territorial expansion, and rich cultural flourishing.

- ◆ **Second Intermediate Period - c. 1650 BCE - 1550 BCE**

Characterized by the invasion and rule of the Hyksos, a foreign dynasty originating from Western Asia, which led to significant shifts in power dynamics within Egypt.

- **New Kingdom - c. 1550 BCE - 1070 BCE**

Following the expulsion of the Hyksos, this period marked the golden age of ancient Egypt, where the civilization reached the height of its power and influence. This era included the reigns of famous pharaohs such as Hatshepsut, Thutmose III, Akhenaten, and Tutankhamun.

- **Third Intermediate Period - c. 1070 BCE - 664 BCE**

Marked by political fragmentation, foreign invasions, and internal conflicts, this period saw the decline of central authority and the emergence of regional power centers throughout Egypt.

- **Late Period and Ptolemaic Period - c. 664 BCE - 30 BCE**

During this time, Egypt was dominated by foreign powers such as the Assyrians, Persians, and Greeks, leading to significant changes in governance and culture. The period culminated in Alexander the Great's conquest and the establishment of the Ptolemaic dynasty, which ruled Egypt until the death of Cleopatra VII and the subsequent annexation of Egypt by the Roman Empire.

- **Roman and Early Christian Period - c. 30 BCE - 4th century CE**

In this period, temples continued to be built in the traditional Egyptian style, and the ancient religion was still practiced. However, the influence of Roman gods and imperial cults began to merge with local beliefs. With the spread of Christianity in the 1st century CE, Egypt became a center of early Christianity. By the 4th century, the Coptic Church had emerged, developing its unique liturgy and language (Coptic, derived from ancient Egyptian). Many temples were converted into churches, and Christian monasticism thrived in the Egyptian deserts, preserving aspects of Egyptian heritage through language, art, and religious traditions despite the suppression of ancient practices.

◆ Islamic Conquest and Influence - 641 CE onward

In 641 CE, the Arab Muslims conquered Egypt, initiating the Islamic influence in the region. The introduction of Islam and the Arabic language significantly transformed Egyptian culture, with Arabic eventually replacing Coptic as the dominant language and Islamic architectural and artistic styles becoming prevalent. Despite these changes, elements of ancient Egyptian culture survived and were often integrated into Islamic traditions, such as the reverence for sacred spaces and the continuation of certain folk practices rooted in ancient customs.

Throughout the medieval period and into the modern era, Egypt continued to be a melting pot of cultural influences. The legacy of ancient Egypt persisted in the form of monumental architecture, such as the Pyramids and temples, which remained symbols of the country's rich history. European explorers and scholars in the 18th and 19th centuries reignited global interest in ancient Egypt, leading to the birth of Egyptology. This renewed interest helped the preservation and study of ancient Egyptian artifacts and texts, further integrating Egypt's ancient legacy into global culture.

Today, the cultural legacy of ancient Egypt is a source of national pride and a major aspect of Egypt's identity and tourism industry. Museums, archaeological sites, and ongoing excavations continue to reveal the depth and richness of ancient Egyptian civilization. Modern Egyptians, particularly the Coptic community, still maintain traditions that reflect their ancient heritage. Moreover, the influence of ancient Egypt is evident in various fields, including art, architecture, literature, and popular culture worldwide.

The influence of ancient Egyptian civilization persists in the cultural identity of Egypt and the fascination it evokes worldwide, serving as a bridge between our ancient past, contemporary life, and possibly our future!

"Egypt is not a civilization that was built overnight. It is the work of 3,000 years of constant development, yet many of its secrets remain hidden in the sands, waiting to be unearthed."

— Zahi Hawas

2. Unraveling the Secrets of Ancient Egypt

Are you ready to explore some of the most well-known aspects of ancient Egyptian achievements and reflect on their place within the world of ancient wisdom? It may be useful to bear in mind that there are still many unresolved mysteries - mysteries surrounding how this civilization managed to accomplish so much.

Apart from the construction of magnificent structures such as the Great Pyramid of Giza, we must not forget the enigma of Egyptian religious and metaphysical beliefs. Their elaborate funerary practices, particularly the process of mummification, were designed to preserve the body for an afterlife that they believed was as real as the physical one. What drove their obsession with immortality, and how did they come to possess such intricate knowledge of anatomy and preservation?

Moreover, the hieroglyphs they left behind, including the still undeciphered symbols and passages, hint at a deeper spiritual and esoteric knowledge, the full extent of which we may never fully grasp.

These unresolved mysteries contribute to the allure of ancient Egypt. Were they merely a highly advanced ancient civilization, or custodians of a much older, hidden wisdom — perhaps remnants of even earlier forgotten cultures like the mythical Atlantis? Each discovery made in the sands of Egypt seems to raise more questions than it answers, keeping the fascination with this ancient civilization alive across the ages.

We know that the inhabitants of ancient Egypt had very sophisticated religious practices and included intricate symbolism in their architecture. These elements went on to influence later civilizations such as the Greeks and Romans. Ancient Egypt has, therefore, long been associated with the Mysteries (a series of ancient religious and spiritual rituals that were practiced in various cultures). These secretive rites involved initiations into esoteric knowledge and spiritual experiences that were intended to bring the initiate into closer communion with the divine, granting them insights into the nature of life, death, and the afterlife. The term "Mysteries" actually comes from the Greek word "mysteria," meaning "secret rites."

From very early on, the ancient Egyptians practiced a highly developed system of religion and spirituality, rich in symbolism and ritual. They believed in the afterlife, the soul's journey, and the concept of divine order (Ma'at). Their religious practices, including the elaborate rituals surrounding death and the afterlife, were seen as holding secret knowledge about the cosmos and the human soul. This gave rise to the idea that Egypt was the keeper of some kind of hidden, esoteric wisdom.

Their grand temples of Luxor, Karnak, and Philae, were not just places of worship but also centers for initiation into the mysteries of the divine. It is believed that in these temples, sacred rituals were carried out by priests and initiates, who underwent processes of spiritual transformation. The secrecy surrounding these rites contributed to the mystique of Egypt as a land of hidden wisdom and mystical traditions.

Egypt's influence on Greek philosophy and the establish-ment of Mystery Schools in Greece and later Rome further strengthened its association with the Mysteries. Many early Greek philosophers, including Pythagoras, Plato, and Solon, are said to have traveled to Egypt to study under the priests and gain knowledge of their secret doctrines. This knowledge gradually influenced the development of Greek philosophical thought and was integrated into the Greek Mystery traditions, such as those at Eleusis. For the Greeks, Egypt was a source of much of their wisdom, in-cluding mathematics, astronomy, and spiritual knowledge.

During the Hellenistic period, especially under the Ptolemaic dynasty, Egypt's association with the Mysteries deepened as Greek and Egyptian religious practices blended. The cult of Isis, originally an Egyptian goddess, became one of the most popular mystery religions in the Roman Empire, spreading Egyptian mystical traditions across the Mediterranean world. The mysteries of Isis, which promised salvation and eternal life, attracted many followers and cemented Egypt's reputation as a land of profound spiritual knowledge.

The grandeur and enigmatic nature of Egypt's monuments, particularly the pyramids and the Sphinx, contributed to its association with the Mysteries. The pyramids were seen as symbols of the soul's ascent to the divine, and their precise alignments with celestial bodies hinted at a deep understanding of the universe's hidden laws. The Sphinx, with its inscrutable expression and leonine form, became a symbol of mystery itself, guarding the secrets of the ancient world.

Today, we continue to be in awe of the Great Pyramids and the spectacular Sphinx, as well as being familiar with mysterious myths surrounding the curse of the Pharaohs. I want to probe deeper into these three topics, going beyond what is common knowledge, to discover the uncommon.

Some interesting theories

Let me begin by pointing out that there are many theories surrounding the enigmas of ancient Egypt, some of which are rather outlandish in their assertions, and others attempting to explain timeless puzzles that remain unsolved.

One of the most famous alternative theories suggests that the pyramids, particularly the Great Pyramid of Giza, were built with the assistance of extraterrestrial beings. Another alternative theory suggests that they were not tombs at all, but highly advanced energy generators or power plants. Nikola Tesla, the renowned inventor and electrical engineer, had some intriguing thoughts about the pyramids in the context of his theories about energy and electromagnetic fields. He was intrigued by the shape of the pyramids and believed they had a unique ability to concentrate energy. Conducting experiments involving the pyramid shapes and their alignment to the Earth's magnetic field, he was curious about their potential to harness natural forces.

Other theorists claim that the pyramids were built as enormous beacons or markers for ancient space travelers, serving as navigational aids for extraterrestrial visitors. One of the leading voices associated with the theory is Erich von Däniken, a Swiss author best known for his controversial book *Chariots of the Gods?* (1968). In it, he argues that many ancient structures, including the Egyptian pyramids, were influenced or directly constructed by extraterrestrial beings, stoking the *Ancient Astronauts* theory.

Regarding the Great Sphinx of Giza, theories have been put forward that it serves as a "stargate," or a portal to other dimensions or parts of the universe. Drunvalo Melchizedek, a New Age author and spiritual teacher known for his work on sacred geometry, consciousness, and ancient mysteries, is just one of the influential thinkers amongst the alternative spirituality community interested in discovering the hidden purpose of these ancient structures.

There is also a belief among some alternative theorists that the Great Sphinx guards a hidden chamber or Hall of Records beneath it, contains ancient knowledge or even evidence of extraterrestrial contact. Edgar Cayce was an American clairvoyant who claimed to have the ability to access information from a higher consciousness or "Akashic Records" while in a trance state. In his readings, Cayce frequently mentioned the existence of a Hall of Records buried beneath the Great Sphinx of Giza, describing it as containing ancient knowledge. He went on to claim it included details about the lost civilization of Atlantis, human origins, and possibly even records of extraterrestrial contact.

One more extreme claim is that the gods of ancient Egypt were actual extraterrestrial beings who were revered as gods due to their advanced technology and abilities. This theory is often tied to the idea that the gods came from another planet and imparted their knowledge to the Egyptians, influencing their culture, architecture, and religious practices. Zecharia Sitchin adds his influential voice to this theory in his books, *The 12th Planet* and *The Earth Chronicles* series, popularizing the idea that extraterrestrial beings played a significant role in shaping human history.

Form a layman's perspective, the main questions that usually pop into our minds when we seek to discover the secrets of ancient Egypt are:

- What exactly was the purpose of the pyramids, other than that of being resting places for deceased Egyptian royalty?

- What secrets does the Sphinx hold beneath its great paws, and why has so much research into its significance failed to come up with answers?

- How are we to interpret the myth surrounding the curse of the Pharaohs - is it really a mysterious hex or a sheer coincidence?

These are the three topics of interest we will explore in this Chapter, taking the opportunity to flesh out some of ancient Egypt's unsolved mysteries along the way.

Let's begin with the Great Pyramids of Giza, admired for their architectural and engineering excellence, as well as for their mysterious links to possible extraterrestrial life. Although Giza is not the site of the first pyramids to be built, it has attracted attention due to its remarkable precision in alignment with the cardinal points and its sophisticated construction techniques, which some believe are beyond the capabilities of the ancient Egyptians.

"The skies have been the mover of [man's] science for millennia, they are his hopes and dreams of tomorrow; nowhere is the vision of the first men who carved their thoughts on stone so fully displayed as in the tombs of earliest Egypt."

— Jane B. Sellers

2.1 The Great Pyramids: Architectural Marvels or Extraterrestrial Blueprints?

The Three Great Pyramids of Giza stand as a monumental legacy of ancient Egypt's engineering prowess. Khufu, the son of Sneferu, initiated the construction of the Great Pyramid around 2580 BCE on the Giza plateau. This massive structure, originally 147 meters tall, was a feat of precision engineering, aligning almost perfectly with the cardinal points and constructed from approximately 2.5 million limestone blocks. Sir Flinders Petrie's 19th-century survey highlighted its near-perfect alignment, showcasing an accuracy still unmatched in modern architecture. The Great Pyramid's casing stones, once polished and gleaming, further demonstrated the builders' exceptional skill.

Following Khufu, his successors Khafra and Menkaura erected two additional pyramids at Giza, each contributing to the grandeur of the site. Khafra's pyramid, slightly shorter but situated on a higher point, appears taller, while Menkaura's, though smaller, still impresses. Nearby, the pyramid of Khufu's son, Djedefra, and another possibly planned by Nebka, have largely vanished, leaving only traces of their former majesty. The precision with which these pyramids were aligned suggests the use of stellar observations, hinting at a deep astronomical and religious significance that guided their construction. The decline in pyramid building after the Fourth Dynasty marks a mysterious end to this era of architectural brilliance.

It is a commonly-held belief that the great pyramids were chiefly built as burial shrines for the pharaohs and their descendents, earthly resting places that assured their safe passage to the afterlife. Nonetheless, countless theories and hypotheses abound surrounding the enigma of the pyramids and their true purpose. Definitive answers to all these questions have not been conclusively established and they continue to captivate the imagination and curiosity of researchers, archaeologists, and the general public alike.

Not surprisingly, the idea of the pyramids being primarily resting places has prevailed through the ages due to the compelling evidence. Ancient Egyptian texts, such as the Pyramid Texts, which are inscribed on the inner walls of these monumental structures, explicitly describe the pyramids as tombs for those laid within. These texts include many spells, prayers, and instructions for the afterlife, indicating their funerary purpose.

When sarcophagi were initially found within the burial chambers of pyramids, such as Khufu, Khafre, and Menkaure at Giza, the idea was reinforced that these structures were designed to house the remains of deceased rulers. Many items typically associated with burials, including canopic jars (used to store the internal organs of the deceased), funerary boats, and offerings meant to accompany the pharaoh into the afterlife were also found within the pyramid chambers.

The pyramids were part of larger mortuary complexes that included temples and causeways designed for rituals and ceremonies associated with death and the afterlife. These complexes further supported the idea that the pyramids were royal tombs. The early historian Herodotus wrote about the pyramids in the 5th century BCE when he visited Egypt, describing them as tombs for the kings. Their design and layout also shared similarities with other known Egyptian tombs, such as mastabas (bench-like tombs with flat roofs), which were predecessors to the pyramids and served similar funerary functions.

Burial Before the Pyramids

In predynastic Egypt, the dead were buried in simple desert pits, with bodies placed in a fetal position, seemingly prepared for rebirth in the afterlife. The dry conditions of the western desert naturally mummified the bodies, likely more by accident than design. However, these simple graves were vulnerable to scavenging by jackals and wild dogs, making them easy targets for tomb robbers. In response, during the First Dynasty, Egyptians began constructing tombs with protective superstructures made of mud bricks and stone, the mastabas.

These massive, rectangular structures with flat roofs were built to safeguard the burial pit and preserve the corpse. Initially used for both kings and nobility, mastabas were eventually reserved for the elite, as dead kings were honored with more grandiose "Mansions of Eternity" from the Third Dynasty onward. Mastabas were often more elaborate than the homes of the living, reflecting the importance placed on the afterlife.

The Stone Pyramids

During the Third Dynasty, Egypt introduced step-pyramids, which, despite their name, are more accurately described as stepped towers. The most famous of these is King Zoser's step-pyramid at Saqqara, the largest of its kind and the first structure in Egypt built with stone masonry — quarried and carefully cut stones rather than roughly stacked ones. This architectural leap is credited to Imhotep, a brilliant priest-architect and Zoser's vizier. Imhotep, later deified and associated with the Greek god of medicine, Asclepius, was also a high priest and the "Chief of the Observers."

The step-pyramid of Zoser, standing sixty meters high with a rectangular base, was more than just a royal tomb — it was a powerful statement of religious belief and artistic achievement. Its ziggurat-like structure, with six steps leading to a seventh platform, likely symbolized the soul's journey through the planetary spheres after death. Visible from Memphis and the surrounding Nile Valley, this monument served as a constant reminder that life was a preparation for the afterlife. Such early structures laid the foundation for the true pyramids of the Fourth Dynasty, including the iconic pyramids of Giza, which may have been influenced by Imhotep's visionary designs.

Pyramids as Mortuary Monuments

Going back to early Egypt, we can understand the origin of the pyramid as a mortuary monument, reflecting the religious beliefs of the ancient Egyptians, who saw life continuing after death in the Hereafter. Initially, there were local cults, but over time, with the unification of Upper and Lower Egypt under a ruling noble class, religious beliefs began to merge.

The evolved Egyptian religion comprised two primary systems: the solar system centered in Heliopolis with Ra as the supreme god, and the Osirian system, which focused on Osiris, the god of the dead, from Busiris and Abydos. The solar system likely predates the Osirian and is rooted in creation myths where the sun god Atum rises from the chaotic primeval ocean Nun, initiating creation. This cosmogony would appear to reflect the natural phenomenon of the Nile's annual inundation and the subsequent emergence of land, symbolized by the primordial hill ben-ben.

The pyramidal shape of the Egyptian pyramids held great significance in ancient Egypt. It represented the sacred mound, a symbol of creation and rebirth. The shape was believed to connect the earthly realm with the heavens, serving as a bridge between the mortal world and the divine realm. The pyramids were constructed with immense precision, aligning with the cardinal directions and incorporating astronomical alignments. The internal structure of the pyramids included various chambers and passages, designed to protect the pharaoh's body and belongings and facilitate his journey to the afterlife. The pyramids were not only architectural marvels but also powerful symbols of the pharaoh's divine authority and his eternal existence beyond death. The pyramidal shape was chosen for its religious significance, stability, and durability, and it represented the pharaoh's journey to the afterlife and his ascent to join the gods.

The Great Mystery

Looking at the Great Pyramid of Giza, the only thing that was actually found in this colossal structure when first breached in 820 BCE was a lidless granite sarcophagus in the so-called King's Chamber. Exploration along the dark tunnels and galleries revealed a system of three chambers, all of which were empty, much to the disappointment of the treasure seekers. Was this mammoth structure really only built to house one coffin for a dead king?

By the time of Tutankhamun (c. 1300 BCE), the Giza pyramids were over a thousand years old, and the memory of who built them and why had been lost. The Greeks and Romans, who occupied Egypt from the fourth century BCE to the seventh century CE, showed little interest in these monuments. However, the Greek historian Herodotus, sought to explain their origins and purpose in *The Histories*. It would appear that much of his account is actually a mix of personal bias, local gossip, and mythology – not really a robust historical description at all.

Since then, speculation continued to shroud the hidden chambers within, with many believing that a secret room lies undiscovered. Generations of Egyptologists and amateurs have searched for it, employing methods ranging from dynamite to x-rays, but, so far, without success. While recent archaeological findings continue to reveal new insights into the Great Pyramid's construction, its real purpose still remains a matter of debate.

Initiatives such as the ScanPyramids project have recently made use of advanced technologies to identify several anomalies within the Great Pyramid. Through muon tomography and infrared thermography, a large void known as the "Big Void" found above the Grand Gallery suggests the presence of unknown chambers or structural spaces. Small camera and sensor-equipped robots have explored previously inaccessible shafts and passageways, uncovering strange features, such as copper handles and door-like structures, suggesting that these hidden chambers might still contain mysterious secrets.

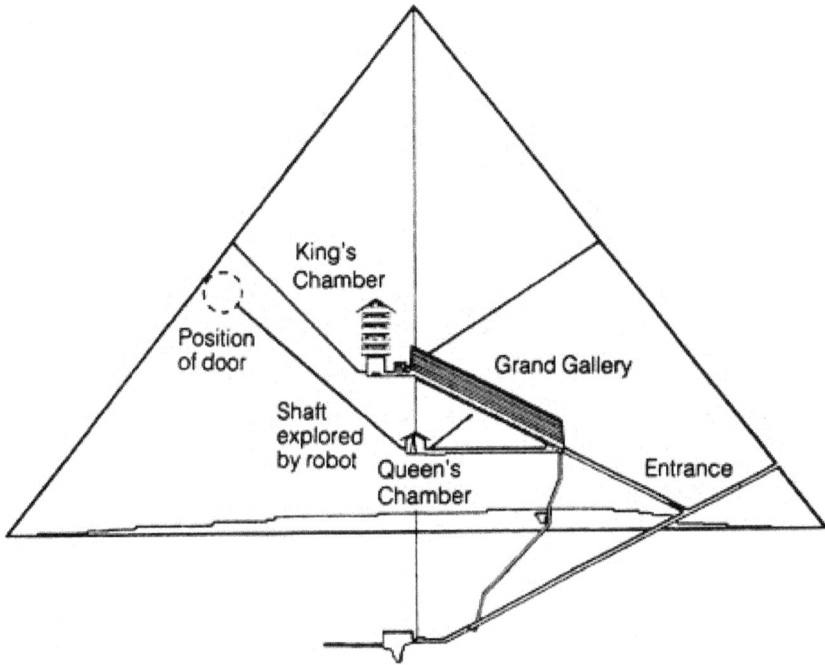

King's Chamber

Position of door

Shaft explored by robot

Queen's Chamber

Grand Gallery

Entrance

More rigorous studies on the alignment of the pyramid have also reaffirmed its precise orientation to the cardinal points and its alignment with specific stars, such as the constellation Orion. This is clear evidence of the pyramid's role in ancient Egyptian cosmology and its symbolic representation of the pharaoh's journey to the afterlife.

Through such recent investigations, It is becoming even more obvious that the pyramids were not just designed to entomb Egyptian leaders, but were much more than that.

The Power of the Pyramid

It is remarkable to note that pyramids appear in various forms across the globe. What is so special about the pyramid shape that led it to be formed by distinctly unique civilizations, from South America and Asia to Africa? In India, for example, we have temples constructed in the shape of pyramids, which is a strange occurrence if we assume that architectural styles and cultural practices were largely isolated. How did these triangular-like structures appear in different parts of the world in similar eras when there was no cross-cultural communication back then? Below are some examples of pyramid-like structures scattered around the world that continue to challenge our perceptions of what we 'thought' we knew about history:

1. Mesopotamian Ziggurats

Located in present-day Iraq and Iran (ancient Mesopotamia), Ziggurats are terraced step pyramids made of mud-brick and baked brick. Unlike the smooth-sided pyramids of Giza, these ziggurats have a more rectangular, stepped design with multiple levels or tiers. Ziggurats served as temples and were dedicated to the gods of the Sumerian, Akkadian, Assyrian, and Babylonian cultures, the most famous of which is the Great Ziggurat of Ur, dedicated to the moon god Nanna.

2. Mesoamerican Pyramids

Located in present-day Mexico, Guatemala, Belize, Honduras, and El Salvador, these pyramids are characterized by their step-like structure with a flat top, where temples were often built. They were constructed by pre-Columbian civilizations such as the Maya, Aztec, and Olmec, for the purpose of religious ceremonies, rituals, sacrifices, and astronomical observations. Notable examples include the Pyramid of the Sun and Pyramid of the Moon at Teotihuacan, the Pyramid of Kukulcán at Chichén Itzá, and the Temple of the Inscriptions at Palenque.

3. Sudanese (Nubian) Pyramids

These pyramid forms can be found in present-day Sudan, particularly in the region of Nubia. They are smaller and steeper than their Egyptian counterparts, having a narrower base and rising to a sharp point, with many constructed from stone blocks. The Vubia pyramids were built as tombs for the kings and queens of the ancient Kingdom of Kush, which was heavily influenced by Egyptian culture. The site of Meroë alone is home to over 200 Nubian pyramids.

4. Chinese Pyramids

In China, several pyramid-like tombs and structures have been discovered, particularly in the Shaanxi province. These are burial mounds rather than true pyramids but have a pyramid-like appearance. The Tomb of the General near Xian and the Mausoleum of the Western Han Dynasty are prominent examples. These tombs are often large and mound-shaped, built to house the remains of emperors and high-ranking officials.

5. North American Mounds

In North America, particularly in the Mississippi River Valley, mound structures with pyramid-like shapes have been found. Again, these are not true pyramids but are stepped or tiered earthen mounds.The largest and most well-known are the Cahokia Mounds in Illinois, including Monk's Mound, which were used for ceremonial purposes.

6. Japanese Pyramids

Some ancient structures resembling pyramids can be found in Japan, though they are primarily burial mounds or kofun (ancient tombs). The Daisen Kofun is the largest ancient tomb in Japan with a pyramid-like shape, offering great insights into early Japanese history and culture.

7. European Pyramids

While not as common, some ancient structures in Europe show pyramid-like characteristics. These are often burial mounds or ceremonial sites, such as the Giant's Grave in Ireland and the Pyramid of Giza-style mounds in the UK. Both are good examples of where pyramid-like forms have been found.

8. Australian Pyramids

In Australia, some natural formations that have been interpreted as pyramid-like by enthusiasts have been found, although these are not man-made. For example, the Pyramid Hill in Victoria and Mount Arapiles are often cited in discussions about pyramid-like natural formations, though they are geological rather than architectural.

9. Indian Pyramids

As I mentioned above, India also has structures with pyramid-like shapes, particularly in the context of ancient temples and monuments. The Sun Temple in Konark, for example, features a pyramid-like shape, and some ancient temples in South India have stepped pyramid structures.

10. Polynesian Structures

Ancient structures and burial mounds resembling pyramids can also be found in Polynesia. The Ahu of Easter Island, although not a true pyramid, features stepped platforms with a similar visual impact.

These pyramid-like structures across different cultures and regions illustrate the widespread fascination with this geometric form and its application in various architectural and ceremonial contexts. The similarity in shape often points to a universal symbolic significance or practical considerations in the construction and design of these ancient structures.

Such resemblances strongly suggest that pyramid shapes have held universal symbolic significance for thousands of years, or it could point to a deeper, perhaps unrecognized, commonality in human ingenuity and religious expression across different civilizations. Especially in esoteric traditions, including Freemasonry, the Rosicrucians, and the Illuminati, the geometric form and symbolic meanings have been central to numerous secret rites and rituals. Here's a deeper look at some aspects of those beliefs in relation to the pyramid form:

1. Ancient Rituals:

Across cultures, the pyramid has been a focal point for rituals and sacred practices. In ancient Egypt, the pyramids were thought to be aligned with celestial bodies and used in rites that connected the pharaohs with the divine.

2. Geometric Perfection and Unity:

The pyramid is often seen as a symbol of geometric perfection and unity. Its base, being a polygon, supports the apex, creating a strong, stable structure. This stability and the convergence of lines towards a single point symbolize spiritual ascent, enlightenment, and the journey towards a higher state of consciousness.

3. Spiritual and Esoteric Significance:

In many esoteric traditions, the pyramid represents a bridge between the earthly and the divine, a conduit for spiritual energy. The shape's alignment with the cardinal points and celestial bodies is thought to connect the material world with higher realms.

4. Symbolism in Rituals:

Freemasonry incorporates various symbols into its rituals, and the pyramid is often used symbolically. The all-seeing eye, often depicted within a pyramid, is a prominent symbol in Freemasonry, representing divine providence and enlightenment. The image of the eye within the pyramid appears on the Great Seal of the United States, reflecting Masonic influence on national symbolism.

5. Historical Influence:

Freemasonry's use of the pyramid and other symbols reflects its roots in ancient mystery schools and eso-teric traditions. The pyramid's symbolism is tied to the broader concept of seeking hidden knowledge and personal transformation.

6. Mystical and Alchemical Symbolism:

The Rosicrucians, a mystical and philosophical secret society, use the pyramid as a symbol of hidden knowledge and spiritual insight. In Rosicrucian lore, the pyramid represents the attainment of higher wisdom and the process of alchemical transformation.

7. Connection to Ancient Knowledge:

Rosicrucianism often links itself to ancient Egyptian and other esoteric traditions that revered the pyramid as a powerful symbol of cosmic order and divine knowledge. The pyramid shape is believed to be a key to unlocking the mysteries of the universe.

8. Conspiracy and Symbolism:

The Illuminati, a purported secret society often associated with various conspiracy theories, is said to use the pyramid symbol to represent its supposed control and influence over global affairs. The "Eye of Providence," often depicted within a pyramid, is associated with the Illuminati in conspiracy theories as a symbol of surveillance and hidden power.

9. Philosophical and Esoteric Dimensions:

In more philosophical contexts, the pyramid represents enlightenment and the illumination of the mind. The Illuminati's use of such symbols is often interpreted as reflecting their alleged pursuit of higher knowledge and secret truths.

10. Modern Esoteric Traditions:

Modern esoteric traditions, including various New Age movements, continue to use the pyramid shape as a symbol of spiritual ascent and connection to higher realms. The geometric precision and symbolic depth of the pyramid make it a powerful tool in rituals aimed at personal and collective transformation.

11. Pyramid Energy:

The concept of "pyramid power" gained prominence in the 20th century, particularly through the work of French architect and researcher Antoine Bovis. Bovis claimed that pyramids, due to their geometric shape, could generate a field of energy that had beneficial effects on living organisms. His experiments included placing food and other items inside small pyramid models, where he observed that they seemed to last longer and decay more slowly.

12. Energy Amplification:

Proponents of pyramid power suggest that the pyramid shape can focus and amplify natural energies. They believe that the geometric configuration of a pyramid — its base and apex — aligns with Earth's magnetic field and cosmic energies, thereby enhancing healing and vitality. This is thought to be due to the pyramid's ability to concentrate and direct energy in a specific way.

13. Healing and Meditation:

In the realm of alternative healing and meditation practices, pyramids are sometimes used as tools to enhance meditation experiences and promote well-being. It is believed that sitting or lying within a pyramid structure can help balance energy fields, promote relaxation, and enhance the body's natural healing processes. Some holistic practitioners use pyramids made of materials like quartz crystal, believing that the combination of the pyramid shape and crystal properties can magnify energy and facilitate healing.

The pyramid's enduring significance across these diverse traditions underscores its powerful symbolism as a connector between the material and spiritual worlds, a beacon of enlightenment, and a vessel of hidden knowledge. It is no wonder, then, that the majestic pyramids of ancient Egypt continue to beguile us with their enormity and tangible power.

The Evolution of Egyptology

While countless theories about the pyramids' mystical purpose — some quite literally "out of this world" — capture our imagination, it's important to put that into context of how the field of Egyptology has evolved over the years. What began as a focus on evidence-based interpretations has grown to include cognitive and narrative approaches, revealing just how much has changed since the days of early excavators.

In the 19th and early 20th centuries, pioneers like Jean-François Champollion, who famously deciphered the Rosetta Stone, concentrated on discovering, cataloging, and translating hieroglyphic texts. Meanwhile, explorers such as Giovanni Battista Belzoni were more concerned with uncovering tombs and artifacts. Although they meticulously documented Egypt's vast wealth of artifacts, monuments, and texts, little attention was paid to the cultural and social contexts of the time. The early approach was largely about establishing timelines, understanding dynastic successions, and detailing the material culture of ancient Egypt.

By the mid-20th century, the introduction of scientific methods sparked a new wave of processual approaches in Egyptology. This era brought innovations like radiocarbon dating, statistical analysis, and environmental archaeology, which allowed Egyptologists to delve into the processes behind Egypt's development. Scholars began analyzing the functional aspects of Egyptian civilization — its agricultural practices, settlement patterns, and economic systems — while remaining firmly rooted in empirical data and a scientific lens.

As the 20th century progressed, Egyptologists like Barry Kemp and Lynn Meskell shifted focus towards the symbolic and ideological facets of Egyptian culture. They explored the religious beliefs, rituals, and symbolic meanings embedded in art and architecture, and started paying closer attention to the experiences and agency of individuals. Scholars began to consider the lives of ordinary Egyptians, their beliefs, and their roles within the broader societal framework. This period also saw a greater emphasis on the context in which artifacts were found, with archaeologists becoming more reflective about their interpretations, acknowledging modern biases and the influence of contemporary cultural perspectives on their work.

Today, Egyptology is a highly interdisciplinary field, drawing from anthropology, art history, linguistics, and more. Cutting-edge techniques like DNA analysis, isotopic studies, and advanced imaging technologies are now essential tools for gaining new insights into ancient Egyptian life. Digital technologies, including 3D modeling and Geographic Information Systems (GIS), enable Egyptologists to create detailed reconstructions of ancient sites and visualize their spatial relationships, blending empirical data with a more experiential understanding.

These advancements have opened up new ways of understanding the knowledge and culture of the ancient Egyptians, allowing us to develop fresher theories about their motivations and beliefs. While empirical data remains crucial, these interpretive approaches offer a richer, more nuanced understanding of the past. With this broader perspective, it becomes increasingly clear that the pyramids were likely built not just to glorify dead pharaohs, but also as profound affirmations of the Egyptians' religious convictions.

Secrets and Mysteries

The Egyptians were an exceptionally reserved people and closely guarded the inner mysteries of their religion, which were known only to a select few initiates. Since these chosen individuals oversaw the construction of the pyramids, it is not surprising that their true motives remain largely unknown to us.

Although we no longer have access to those ancient minds, nor are we able to ascertain the source of their knowledge, several notable voices in the world of alternative Egyptology offer interesting ideas about the real purpose of the Great Pyramids, and we will take a look at some of those ideas below.

Robert Bauval is an Egyptian-born author and lecturer widely recognized for his Orion Correlation Theory, which he introduced in the 1994 book *The Orion Mystery,* co-authored with Adrian Gilbert. Bauval's background in engineering and his lifelong fascination with ancient Egypt fueled his exploration into the astronomical alignments of the Giza pyramids. The Orion Correlation Theory suggests that the three pyramids of Giza were deliberately aligned with the three stars of Orion's Belt, reflecting the ancient Egyptians' deep connection with the cosmos, particularly their belief that the pharaohs were destined to join the gods in the afterlife.

Bauval's work opened up new interpretations of ancient Egyptian architecture, suggesting that the placement of the pyramids was not merely for show or religious purposes but was part of a grander cosmic design. His theory has led to broader speculations about the purpose of the pyramids, with some theorists suggesting that these structures could have functioned as stargates or portals due to their precise celestial alignments. Bauval himself remains somewhat agnostic on the more extreme interpretations of his theory, focusing instead on the profound symbolism of the pyramids' alignment and their significance in the context of ancient Egyptian cosmology.

Graham Hancock is a British author and journalist whose works have often been at the center of controversy due to his unconventional theories about ancient civilizations. Hancock's 1995 book *Fingerprints of the Gods* catapulted him into the spotlight by proposing that ancient monuments like the pyramids are remnants of a lost advanced civilization that predated known history. Hancock's theory challenges the traditional narrative of human history, suggesting that these structures were not only built with advanced knowledge of astronomy and engineering but also served purposes that go beyond the mere burial of kings or religious rituals.

Hancock extends his theory to propose that these ancient sites, including the pyramids, might have been designed with a sophisticated understanding of the cosmos, possibly serving as stargates or portals that facilitated communication with other realms or beings. In his 2015 sequel, *Magicians of the Gods,* Hancock revisits these ideas, arguing that ancient myths and legends from around the world point to a forgotten episode in human history when such advanced knowledge was widespread. His work has been influential in popularizing alternative history and continues to inspire debates about the true nature and purpose of ancient monuments.

David Hatcher Childress is an author and publisher known for his extensive writings on ancient mysteries, lost civilizations, and unexplained phenomena. Through his books, particularly *Technology of the Gods* and the *Lost Cities Series,* Childress explores the possibility that ancient civilizations possessed advanced technology that has since been lost or forgotten. He often speculates that the pyramids and other megalithic structures could have been part of a global network of stargates or portals, enabling communication or travel between different parts of the world or even other dimensions.

Childress is a prolific figure in the realm of alternative archaeology, frequently combining evidence from diverse cultures to suggest that our understanding of ancient technology is incomplete. Extraterrestrial beings or a highly advanced lost civilization may have imparted knowledge to ancient peoples, who then used this knowledge to construct these monumental structures. While mainstream archaeology largely dismisses these claims, Childress's ideas contribute to ongoing debates about the technological capabilities of ancient societies .

William Henry is an investigative mythologist and author who specializes in interpreting ancient myths and texts through a lens that often involves speculative theories about stargates and portals. In works such as *The Healing Sun Code* and *The Secret of Sion*, Henry blends mythology, history, and esoteric knowledge to argue that ancient structures like the pyramids were not just tombs or temples but also technological devices designed to facilitate spiritual and physical travel between different worlds.

Henry's approach is deeply symbolic, viewing the pyramids as metaphysical constructs that reflect a higher understanding of the universe, where physical and spiritual realms intersect. His theories draw on a wide range of sources, including biblical texts, ancient Egyptian mythology, and modern science fiction, to propose that these structures could serve as gateways to other dimensions or planes of existence. Henry's work is part of a broader movement that seeks to uncover hidden meanings in ancient monuments, often connecting them to spiritual and mystical traditions.

Zecharia Sitchin was an author best known for his interpretations of ancient Sumerian texts, particularly those related to the Anunnaki, a group of deities in Mesopotamian mythology. He suggests that the pyramids may have been part of a broader network of structures designed for purposes beyond the comprehension of contemporary historians and archaeologists, possibly functioning as stargates or portals used by the Anunnaki to travel between Earth and their home planet, Nibiru. Even though Sitchin's theories have been widely criticized by scholars, they remain popular among those who subscribe to alternative histories and the idea of ancient astronauts, which we will discuss further in Chapter Six.

In *Manual of the Pyramids: Ancient Time Engineering*, **Warren E. York** proposes a highly speculative theory that ancient pyramids, including the Great Pyramid of Giza, were not merely tombs but sophisticated energy generators and time transportation devices. York suggests that the pyramid shape creates a gravity lensing effect, similar to the gravitational field around celestial bodies like Earth or the Sun, which could slow down time within the pyramid's chambers. He speculates that this effect might even generate a wormhole or photonic portal, potentially explaining the preservation of objects found within the pyramids.

The author also references sightings and photographs of pyramid-shaped objects, including a 2016 image at Giza, as evidence of these phenomena. York proposes that the pyramids might hold the key to understanding ancient civilizations and the possibility of intelligent life in the universe. However, it's important to note that these ideas are not supported by mainstream archaeology or scientific consensus and still remain purely speculative.

Is it possible that the main purpose of the Great Pyramids remains hidden within ancient texts, astronomical alignments, and the pyramids' construction? Are those voices on the periphery of archaeological debate right in their suggestions that the true knowledge about the unique purpose of the Great Pyramids has gotten lost over the centuries? If so, will we ever be able to recover it?

"The alignment of the Sphinx, pyramids, and stars suggests a deeper cosmological connection. It is possible that the Sphinx, facing directly east, was meant to embody the constellation of Leo during the Age of Leo, some 10,500 years ago, linking the monument to an ancient epoch and a forgotten civilization."

— **Robert Bauval**

2.2 The Sphinx: Guardian of Mysteries

Standing at dawn between the colossal paws of the Great Sphinx, bathed in the golden light of the rising sun, is a humbling experience that evokes a sense of timelessness.

Standing at approximately 73.5 meters (241 feet) in length, 20.7 meters (68.5 feet) in height from base to the top of the head, and 19.3 meters (63.5 feet) in width across the shoulders, this feline wonder is carved from a single mass of limestone quarried directly from the Giza Plateau. The Sphinx showcases the ancient Egyptians' advanced engineering and artistic skills, crafting a statue from the rock to reveal the final form. Despite suffering from erosion and vandalism over the centuries, resulting in the loss of its nose and other features, the Sphinx remains formidable.

This monumental statue, believed by many to be far older than the 4,500 years assigned by Egyptologists, might date back to the last Ice Age, challenging our current understanding of ancient civilizations. Geological evidence suggests a much earlier origin, though precise dating remains elusive, with estimates ranging from 15,000 to 5,000 BCE.

The Great Sphinx, with its majestic presence, aligns flawlessly with the cardinal points, marking the sunrise of the spring equinox with astonishing accuracy. This precise alignment suggests that its creators harnessed advanced astronomical knowledge. The Giza plateau, with its sweeping views, may have functioned as an ancient observatory, meticulously crafted to track the movements of the sun and stars.

The Sphinx's exact orientation, along with the Great Pyramid's near-perfect placement at latitude 30 and its alignment with true north-south, underscores the builders' remarkable astronomical expertise. Modern star-mapping technology allows us to revisit the skies over Giza from any point in the past 30,000 years, unveiling celestial alignments that reveal new layers of significance to these ancient structures.

When viewed through this lens, the Sphinx emerges as an equinoctial marker, perfectly aligned with the spring equinox. Fascinatingly, while one might expect an 'Age of Taurus' site to feature bull symbolism, the Sphinx is distinctly leonine. This becomes clearer when considering the 'Age of Leo,' around 10,500 BCE, when the Sphinx's alignment with the vernal equinox perfectly matched its symbolic role.

Simulations from this era show the constellation of Leo rising just before dawn, directly in line with the Sphinx's gaze. At the same time, the three stars of Orion's Belt, culminating at the meridian, align with the layout of the three pyramids.

Together, the Sphinx and the pyramids create a remarkable celestial map, offering a vivid depiction of the stars and highlighting the deep connection between the heavens and the ancient builders of Giza.

The Hall of Records

According to tradition, a hidden "Hall of Records" that lies beneath the Great Sphinx (or possibly within the Great Pyramid itself) holds the ancient wisdom and knowledge of this vanished civilization.

Ancient Egyptian texts and inscriptions, including some enigmatic references in the Pyramid Texts and later writings, hint at hidden knowledge or sacred archives. However, these are often vague and not direct evidence of a specific Hall of Records. The idea of a hidden repository of ancient wisdom has parallels in other cultural myths, meaning it isn't something unique to the Sphinx. For instance, Hermetic texts and later Gnostic writings sometimes refer to hidden knowledge or sacred texts safeguarded from cataclysmic events. Hermes Trismegistus often wrote of divine secrets and cosmic principles that are concealed from ordinary understanding but revealed to the initiated. These concepts resonate with the notion of the Hall of Records as a repository of profound and esoteric knowledge.

The modern concept of the Hall of Records is significantly influenced by Edgar Cayce, the early 20th-century psychic known as the "Sleeping Prophet." Cayce claimed that the Hall of Records was an Atlantean library buried beneath the Sphinx, intended to preserve ancient knowledge after the fall of Atlantis. His readings suggested that the Hall contained records of ancient civilizations, including esoteric and scientific knowledge.

Esoteric traditions often emphasize the existence of hidden or secret knowledge, accessible only to a select few through mystical experiences or advanced spiritual understanding. The Hall of Records fits this theme as it is purported to contain ancient wisdom, accessible only to those who have attained a certain level of enlightenment or insight.

In the 20th and 21st centuries, New Age and occult movements have embraced and expanded upon the idea of the Hall of Records. These interpretations often blend elements of ancient wisdom, astrology, and mysticism, presenting the

Hall as a sacred space that holds the secrets of the universe and human evolution. The Sphinx, in these views, is seen as a guardian of this esoteric knowledge.

Speculative archaeology and fringe theories frequently draw on esoteric ideas to propose hidden chambers or advanced ancient technologies. The Hall of Records is often portrayed in these theories as a physical manifestation of esoteric knowledge, hidden beneath the Sphinx and awaiting redis-covery by those with the right spiritual or intellectual insight.

But why is the Hall of Records thought to be hidden under the Sphinx, of all places?

There are some interesting reasons for this. Firstly, the Sphinx is positioned in a way that aligns with celestial events, such as the equinoxes. This alignment has led to theories that the Sphinx may have been constructed as part of a grander astronomical or symbolic scheme, possibly related to hidden knowledge or sacred records.

Secondly, ancient and medieval sources suggest that import-ant knowledge was hidden to protect it from disasters or to preserve it for future generations. The idea that the Sphinx might conceal such knowledge fits into these broader myths about lost wisdom and ancient civilizations.

Various exploration attempts using technologies like ground-penetrating radar and seismic surveys have identi-fied anomalies in the Sphinx's vicinity. These findings have often been interpreted by some researchers as potential evidence of hidden chambers or tunnels, fueling speculation about a secret repository.

The Hall of Records has also been a popular subject in speculative archaeology and conspiracy theories. Films, books, and documentaries have expanded on the idea, often portraying the Sphinx as guarding an ancient library or repository of lost knowledge, further cementing the notion in the public's imagination.

The allure of the Hall of Records is rooted in its promise of revealing profound, often cosmic, understanding, aligning with esoteric beliefs about the nature of knowledge and the universe. Guarded over for millenia by the colossal, silent Sphinx, we can only wonder if, one day, its secret will be revealed.

In modern times, such theories have been popularized by speculative Freemasonry, esoteric groups, and psychics like the aforementioned Edgar Cayce, who claimed that the Sphinx and the pyramids housed records from Atlantis (more about Atlantis and its links to ancient Egypt in Chapter 3). Cayce's ideas have fueled a substantial New Age industry blending with mainstream Egyptological research, and continue to inspire exploration and debate about the true history and purpose of these monumental structures.

The quest to uncover hidden secrets beneath the Great Sphinx led researchers like Mark Lehner and other pioneers to utilize ground-penetrating radar and advanced remote sensing technologies to investigate anomalies under the Sphinx's bedrock. These early efforts, beginning in 1973-4, were supported by institutions such as Ain Shams University and the Stanford Research Institute (SRI).

In 1977, with funding from the US National Science Foundation, the SRI employed additional techniques, including resistivity measurements and magnetometry. Their findings suggested several anomalies, including a potential tunnel behind the Sphinx's rear paws and other unexplained features in front of the paws.

The ECF/ARE funded further detailed surveys in 1978, focusing on the Sphinx enclosure and the nearby Sphinx Temple. Despite extensive scans and drilling, many of the anomalies turned out to be natural cavities.

The Form of the Sphinx

The form of the Sphinx, with its distinctive combination of a human head and an animal body, is not unique to ancient Egypt, though it is most famously represented by the Great Sphinx of Giza.

Throughout the eras and across various cultures, similar hybrid creatures have appeared in mythologies and art, symbolizing a range of ideas from divine guardianship to the embodiment of complex deities. In ancient Mesopotamia, for example, the Assyrian civilization created winged sphinxes as protective symbols. In Greece, the mythological Sphinx appears with the body of a lion and the head of a woman, acting as a guardian of knowledge and justice. Additionally, in India, the concept of the sphinx-like creature appears in the form of the mythological Narasimha, a deity with a lion's head and human body, symbolizing divine protection.

Let's take a more detailed look at how the form of the Sphinx appeared throughout these cultures:

1. Mesopotamian Mythology

In Mesopotamian mythology, creatures resembling the sphinx were present, such as the Lamassu, which has a human head, the body of a bull or lion, and the wings of an eagle. The Lamassu served as protective deities guarding the entrances of cities and palaces, representing strength, protection, and royalty.

2. Greek Mythology

In Greek mythology, the sphinx was depicted quite differently from its Egyptian counterpart, having the body of a lion, the wings of a bird, and the face of a woman. It is most famously associated with the myth of Oedipus, according to which, the sphinx guarded the city of Thebes and posed a riddle to travelers: "What walks on four legs in the morning, two legs at noon, and three legs in the evening?" Those who failed to answer correctly were devoured by the sphinx. Oedipus eventually solved the riddle (answer: a human), leading to the sphinx's death. In Greek culture, the sphinx continues to symbolize mystery, danger, and enigmatic wisdom.

3. Indian Mythology

In South and Southeast Asian cultures, similar creatures to the sphinx are also found, such as the Purushamriga or Naravirala in Hindu and Jain mythology. These beings have a human head and a lion's body, with the Purushamriga being depicted as a guardian figure, protecting temples and sacred spaces from evil spirits and demons. Another example is the Narasimha, a half-lion, half-man avatar of the Hindu god Vishnu, who is revered for his strength and role as a protector against evil forces.

4. Persian Mythology

In Persian culture, a similar creature is known as the Manticore, which is usually depicted with a human face, a lion's body, and a tail that can either be of a dragon or a scorpion. It shares the sphinx's hybrid form and association with danger and protection, considered to be a fierce creature that could shoot venomous spines to hunt its prey.

5. Ethiopian and Eritrean Mythology

In Ethiopia and Eritrea, the Wagadou is a legendary creature that is similar to the sphinx. It is considered a guardian figure, often depicted in a similar fashion to the Egyptian sphinx with a lion's body and a human head, though with regional artistic differences.

This cross-cultural recurrence of sphinx-like figures underscores what appears to be a shared symbolic language and thematic concerns. Not only in Egypt, but also in cultures throughout the globe, there has been a universal fascination with blending human and animal forms to convey power, mystery, and transcendence.

Is the Sphinx a "Yardang"?

A new study suggests that the Great Sphinx of Giza may not have been entirely sculpted from scratch by the ancient Egyptians but could have been shaped from a naturally occurring rock formation. Published in *Physical Review Fluids* on October 17, 2023,[1] The research from New York University proposes that a "yardang" — a wind-carved ridge — might naturally develop a sphinx-like appearance.

While this theory suggests a natural basis for the Sphinx's shape, it does not diminish the human effort involved. The study acknowledges that the Egyptians would still have needed to refine the monument's distinct facial features and details.

Most Egyptologists and scientists are intrigued by the idea but remain skeptical about the existence of such a naturally sphinx-like formation at Giza.While similar formations may

---

[1] https://link.aps.org/doi/10.1103/PhysRevFluids.8.110503

be observed around the world, none exist that resemble the study's proposed shape.

The enigma of the Sphinx continues to captivate and mystify scholars and enthusiasts alike, leaving us to ponder its true purpose and symbolism. Is it merely a grandiose tribute to a forgotten era, a cosmic marker aligned with celestial precision, or a testament to an ancient civilization's ingenuity? The Sphinx stands as a silent sentinel over the sands of time, its weathered features and enigmatic gaze challenging us to unravel the secrets of its origins and significance.

As new theories emerge and old questions persist, the Sphinx is an enduring symbol of human curiosity and the eternal quest for understanding in the face of the unknown.

"I am skeptical about the curse, but there are forces that we have not yet fully understood. One must tread carefully in these sacred places, where the ancient Egyptians believed their kings were protected even in death."

— Lord Carnarvon

2.3 The Curse of the Pharaohs: Fact or Fiction?

We have all heard of the dreaded curse to fall on anyone who dares to disturb the sacred resting places of the divine pharaohs. The story first began to gain traction when a series of mysterious deaths were reported after the discovery and opening of Tutankhamun's tomb by Howard Carter and his team in 1922. This event was followed by an unsettling incident in which a messenger sent by Carter encountered a cobra — a symbol of Egyptian royalty — at Carter's house, with the cobra having killed Carter's pet canary. Seen as an ominous sign that coincided with the tomb's unveiling, it fueled rumors of a curse.

Lord Carnarvon, who had funded the excavation, subsequently died of an infected mosquito bite, which stirred the public interest as the media fanned the curse rumors. Other notable figures, like Sir Bruce Ingram, faced misfortunes linked to artifacts from the tomb, such as a paperweight made from a mummified hand, which seemed to curse its owner.

Despite popular myths, no curses were actually found engraved in Tutankhamun's tomb, but the fascination with curses intensified after its discovery and the deaths that followed. While some tombs do contain genuine ancient curses inscribed on their walls, they are quite rare and usually aimed at safeguarding the tomb from desecration rather than threatening potential robbers. Most known curses come from private tombs of the Old Kingdom and include warnings against things like impure individuals and violators.

Challenging the Myth

Since the mid-20th century, many authors and documentaries have suggested that the infamous "mummy's curse" might have scientific explanations, such as bacteria, fungi, or radiation. These pathogens, such as Aspergillus, might have thrived in the tombs' dark, humid environments and could cause respiratory or other health issues in those who disturb the tombs.

Some believe that ancient Egyptians may have used toxic substances, like arsenic or other chemicals, in their mummification processes. Over time, these could have become airborne or absorbed through the skin when tombs were opened, potentially leading to health problems.

The power of suggestion and the psychological impact of believing in a curse could even trigger stress-related illnesses, with fear and stress weakening the immune system and making people more susceptible to sickness after entering a 'cursed' tomb.

The tombs themselves might contain environmental hazards unrelated to curses or toxins, such as unstable air quality, high levels of carbon dioxide, or other noxious gasses that could cause dizziness, confusion, or health issues upon prolonged exposure.

On the other hand, the origins and evolution of these curse tales point more to cultural and psychological factors than to actual scientific phenomena. Initially rooted in European fascination with ancient Egypt, these stories evolved from magical warnings to scientific theories, and eventually became a staple of horror fiction. Thus, the "curse" might be more a product of cultural imagination and storytelling than of tangible scientific danger.

Historical accounts of curses often predate the decipherment of hieroglyphs, with early tales reflecting folklore rather than factual curses. Modern literature, such as Jane C. Loudon's 1827 novel entitled *The Mummy!: Or a Tale of the Twenty-Second Century* and *Lost in a Pyramid: Or, The Mummy's Curse*, by Louisa May Alcott in 1869, further popularized the curse legend. Such is the power of these stories, combined with superstitions surrounding the mummified pharaohs, that we continue to be captivated and intrigued. However, up until now, curses are usually seen more as myth than reality.

The Science of Curses

According to a paper published in the Journal of Scientific Exploration[2], modern surveys of Egyptologists working in ancient tombs have uncovered an alarming pattern: many have died from cancers similar to radiation sickness. This parallels reports of unusually high radiation levels in these tombs, which can't be explained by natural background radiation.

Ancient Egyptian texts describe a mysterious "saffron cake," resembling yellowcake uranium, said to possess invisible powers and leave hazardous waste buried in a tomb known

---

[2] Fellowes, R. (2024). The Pharaoh's Curse: New Evidence of Unusual Deaths Associated With Ancient Egyptian Tombs. *Journal of Scientific Exploration*, *38*(1), 41-60. https://doi.org/10.31275/20242855

as the per D'jet, or "house of millions of years." The ancient curse on this tomb warned that anyone who disturbed it would face a mysterious, undiagnosable disease.

The oldest known case of cancer revealed in the paper dates back to ancient Egypt, specifically during the Old Kingdom around 3000-2500 BCE. The case was identified in a young man from ancient Nubia, proving evidence of metastatic carcinoma. Additionally, individuals associated with ancient Egyptian tombs, such as Akerbald, J D and Lepius, K, were also found to have died from cancer. These findings suggest that cancer cases can be traced back to individuals who interacted with the contents of the tombs in ancient times.

What is more disturbing is the theory that ancient Egyptian tombs were used for the deliberate burial of nuclear waste, challenging traditional views of these structures. This hypothesis is supported by evidence of high radiation levels in the tombs, modern-era health impacts on individuals exposed to tomb excavations, textual evidence describing nuclear technology in ancient Egyptian funerary literature, and archaeological records suggesting the storage of hazardous materials in the tombs. The hypothesis also suggests a link between the high incidence of cancer in ancient Egypt, radiation exposure, and the deliberate burial of nuclear waste in these ancient structures.

Howard Carter himself died of Hodgkin's lymphoma, a type of haematopoietic cancer that has been linked to radiation sickness. The question is, what evidence is there to suggest that the ancient Egyptians buried toxic waste in the pyramids in the first place?

Ancient Egyptian funerary texts describe the transformation of Osiris using terms that could be interpreted as related to radioactive materials like yellowcake-235 (U235) and hazardous waste burial. The physical evidence in the form of stoneware pots found in early tombs at Abydos and Saqqara, which were speculated to contain offerings like beer or wine, but actually found to contain substances like mud, plaster, or mortar. This suggests a different purpose for these vessels, possibly for the storage of hazardous materials.

Interestingly, the ancient Egyptian nuclear technology referenced in the above mentioned journal includes the processing of nuclear fuel and the presence of structural and functional characteristics in mastabas consistent with secure, long-term underground storage for nuclear waste.

Additionally, there are descriptions in funerary literature of nuclear-related activities, such as the transformation of Osiris into light and references to 'magic food' processed by nuclear methods. The ancient Egyptians may have handled uranium-based materials, utilizing radioactive substances for rituals, ceremonies, or alchemical practices. They could have managed nuclear waste through burial in underground structures, interacting with materials emitting alpha, beta, and gamma radiation, and implemented practices involving nuclear materials for religious, medical, or technological purposes.

Modern studies have detected unusually high levels of radiation, particularly radon gas, in the tombs, indicating the presence of radioactive materials. In addition, the high incidence of cancer and other health issues among modern-era Egyptologists who worked in these tombs, suggests exposure to radiation.

Could the deliberate and organized practice of dumping toxic waste in the pyramids really be linked to the ancient Egyptians' beliefs in the afterlife, and rituals of transformation. Does this practice, in fact, offer an explanation for the curse of the pharaohs? Only time will tell...

Esoteric Explanations

Being a deeply spiritual people, the Egyptians clearly had a sophisticated understanding of esoteric beliefs. In relation to current views on the "Curse of the Mummies" or "Curse of the Pharaohs" those with an interest in esotericism offer interesting perspectives.

Some esoteric traditions suggest that ancient Egyptian priests or magicians invoked elemental spirits or "elementals" to guard the tombs of pharaohs. These entities, created through ritual magic, were believed to protect the sacred burial sites from intruders, potentially causing harm or death to those who disturbed the tombs. The well-known occultist and author, Dion Fortune, wrote extensively about the use of elementals in magical practices. While not specifically focused on Egyptian tombs, her works, such as *Psychic Self-Defense*, discuss how elementals can be created and used for protection.

Another esoteric view holds that ancient rituals and spells performed during the burial process left behind a powerful energetic residue. This energy, sometimes referred to as a "curse," could manifest as negative forces affecting anyone who disturbs the tomb, leading to illness, misfortune, or even death. Manly P. Hall is a prominent esoteric scholar who has written about the spiritual and energetic aspects of ancient Egyptian practices in his books *The Secret Teachings of All Ages* and *The Kybalion by Three Initiates* (which discusses the concept of vibration and energy that can be applied to understanding energetic residues)

In some esoteric belief systems, the concept of karma plays a role in the curse. It is thought that those who violate sacred spaces or disturb the peace of the dead may suffer karmic consequences, either in this life or future incarnations. The curse is seen as a form of cosmic justice for disrupting the balance between the living and the dead. Helena Blavatsky, one of the founders of the Theosophy movement, wrote about karma and ancient wisdom traditions, including Egyptian, in the influential *The Secret Doctrine*.

Esoteric teachings sometimes describe tomb curses as a form of spiritual warfare, where the spirits of the deceased, along with protective deities or ancestral spirits, actively defend their resting place. These spirits may be angered by the intrusion, resulting in various forms of retribution. The influential occultist Alesiter Crowley mentions such topics and explores various aspects of ritual magic and conflict in his book, *The Equinox*.

Another intriguing esoteric theory proposes that the tombs were designed with spiritual traps intended to ensnare the souls of those who violated the sanctity of the burial. These traps could lead to a form of spiritual bondage or affliction that manifests in the physical world as the curse. Kenneth Grant, a disciple of Aleister Crowley, explores the intersection of Egyptian magic and concepts of spiritual entrapment in his works, *Outside the Circles of Time* and *The Magical Revival,* focusing on Egyptian magic and its implications for the soul.

More modern esoteric interpretation suggests that the tombs might be connected to higher dimensions or realms of existence. The curses could be seen as a way of maintaining the integrity of these connections, with intruders inadvertently triggering negative consequences due to their lack of understanding of these spiritual or interdimensional forces. Graham Hancock has explored ideas related to ancient knowledge and interdimensional forces in his books, *The Message of the Sphinx* (co-written with Robert Bauval), and *Fingerprints of the Gods*, which also suggests advanced knowledge in ancient cultures, hinting at possible interdimensional influences.

These voices represent a range of esoteric and occult perspectives on the curse of the mummies and the spiritual significance of ancient Egyptian tombs. They offer various interpretations of how ancient practices might still exert influence or carry symbolic power today.

Indeed, such esoteric views contribute to the mystique and allure of the mummy's curse, intertwining ancient Egyptian spirituality with modern occult and metaphysical interpretations. They highlight the belief that ancient Egyptians possessed profound knowledge of spiritual forces, and that their burial practices were not only about honoring the dead but also about protecting their sanctity from the profane. Sounds like an interesting interpretation to me!

Ancient Egypt's enduring mystique continues to captivate the world, its legacy woven into the fabric of history and culture across millennia. Even though the sands of time have shifted, the echoes of this extraordinary civilization remain ever-present, inviting us to explore and marvel at the depths of its wisdom, creativity, and resilience.

In a world that often seems distant from its ancient roots, perhaps the allure of Egypt, its monuments, and secrets remind us of the power of the past to inspire the present and shape our future. In that respect, we still have much to learn!

The Oasis to Another Realm

We were backpackers, moving through Egypt on a loosely planned itinerary that had taken us from the crowded markets of Cairo to the quiet temples of Luxor. We'd been following the usual tourist routes at first, but it wasn't long before the allure of something deeper drew us further off the beaten path. The country had that effect on us — its history and mystery seemed to seep into every corner, pulling us toward the less-traveled places. That's how we found ourselves driving deeper into the desert, far from the cities, in search of a small archaeological site we'd heard whispers about from a local guide.

The Land Rover we'd rented wasn't in the best shape, but it was all we could afford. It handled the desert terrain well enough until the sun started to dip, casting the landscape in a golden light. Then, without warning, the engine sputtered and fell silent. No drama, no smoke — just a slow, defeated halt. We tried restarting it, but after a few half-hearted attempts, it was clear that we were stuck.

There wasn't much around, but on the map we could see there was an ancient oasis about half a mile away that occasionally attracted travelers. When we arrived there, we saw that a camp was already set up but, surprisingly, no one was there that night. There were a few palm trees and a pool of water — it was our only option. We grabbed what we could from the car and walked toward it, the sand crunching under our boots, the horizon quickly fading from orange to deep purple as night set in.

By the time we reached the oasis, the desert had fallen into that thick, enveloping silence. The kind that makes you feel like the world has paused. We set up our own tent – a makeshift camp by the water – hoping to wait out the night until we could figure out how to get the car running again in the morning. As I sat there, sipping water and staring at the sky, I wasn't expecting much beyond a quiet night under the stars.

I certainly didn't anticipate anything unusual. But as the night deepened, I couldn't help but notice something odd about the sky. The stars seemed clearer than I'd ever seen, but it wasn't just the clarity. It was the way they aligned — perfectly. The three stars of Orion's Belt were hovering directly above the palm trees, forming a line so sharp and symmetrical that it was hard to believe it was natural.

Then, in the far distance, other stars began to move — not fast like meteors, but with slow, deliberate purpose. They shifted in ways that defied logic, as if the sky itself was alive, responding to some ancient rhythm. A thin, shimmering band of light appeared between them, linking constellations that shouldn't have been connected. I found myself staring, unable to look away, though I couldn't fully grasp what I was seeing.

It wasn't frightening, exactly. In fact, there was something calming about it. Everything felt connected — the stars, the desert, the oasis, and me. Time lost its meaning. I wasn't thinking about the broken car or the long walk ahead of me in the morning. All of that seemed irrelevant compared to the quiet, expansive awareness I was drifting into.

I can't explain it fully. It wasn't a vision or a dream. It felt more like I'd simply shifted into another state of being, like

I was tuning into something that had always been there but that I hadn't noticed before. The alignment of the stars, the activity in the sky — it all seemed to resonate with a deeper part of me, something beyond thought or reason.

By the time dawn broke, the sky had returned to its usual self, the stars fading into the early light as if nothing had happened. The car also magically came back to life!

I couldn't put it into words — not then, and not really now. All I know is that night in the desert felt like I'd glimpsed something far bigger than myself, the car issue seems to have been a blessing, something that still lingers with me in the quiet moments, when I look up at the night sky and wonder about the endless mysteries that lie behind that door.

By the way, let me conclude by saying, I may or may not have been one of the travelers. Perhaps this story was just narrated to me by someone who went on this journey. Nonetheless, it must have been a beautiful experience!

"And so the city of Atlantis, which had been the most powerful and most splendid of all, was swallowed up by the sea and disappeared."

— Plato, Critias 120e

3. The Lost City of Atlantis: Myth or Reality?

The myth of the lost city of Atlantis is an enduring one. For centuries, we have held onto the idea that somewhere in the world, a once dazzling city lies at the bottom of the deep blue sea. But did it ever really exist and, if so, will we ever find it? Where it is located, no one knows, although there are no shortage of theories about its whereabouts. Some provide possible explanations about the fate of this legendary land, while others delve deeper into the mystical significance of the Atlanteans. Which school of thought is right? Should we rely on science to provide us with the answers or be open to other voices that offer alternative, and often beguiling accounts of the fortunes of this fabled land?

Let's begin by looking at the origins of the Atlantis story and then delve into the various interpretations of the civilization's rise and fall.

It was the Greek philosopher Plato who first recorded the existence of the sunken city in his dialogues *"Timaeus"* and *"Critias,"* written around 360 BCE. In these works, he presents Atlantis as a powerful and advanced civilization that existed about 9,000 years before his time that eventually fell out of favor with the gods, leading to its destruction. Stories of whole civilizations disappearing during times of great floods have been embedded in myths across cultures for as long as we can remember, well before the Greek philosopher first made mention of Atlantis. In that sense, the story of this lost city isn't unique, so what makes it so intriguing?

Part of its allure lies in Plato's detailed and enigmatic description of a highly advanced society, with remarkable architecture, technology, and knowledge, which disappeared overnight beneath the sea. The combination of its mysterious location, its advanced culture, and its sudden, tragic downfall has sparked endless speculation, blending history, mythology, and pseudoscience. The ongoing search for Atlantis, whether as a literal place or as a symbol of lost knowledge, continues to captivate imaginations, driving explorers, scholars, and storytellers alike. Its intrigue lies not just in its possible existence, but in what it represents about the fragility of human civilizations and the dangers of unchecked power.

What puzzles most historians is whether or not there is any evidence to support the existence of the lost city. Was Plato retelling events as they had been relayed to him, or was his account more allegorical in nature and not to be taken as historical fact? It is this question that has been the subject of debate for centuries. In this chapter, we will be unlocking the myth, looking at how the search for Atlantis has unfolded, and reflecting on the stories surrounding the legend.

Whatever your views are on the subject, join me as we embark on a quest for the truth, even if that leaves us with more questions than we have now.

"The mystery of Atlantis captivates the imagination because it hints at a lost era of wisdom and technological prowess. It raises questions about human history and the cyclical nature of civilizations."

— Charles Berlitz

3.1 Plato's Account: The Tale of Atlantis

Let's begin by exploring the myth of Atlantis through the account given by Plato.

We can start by saying it is one of the most captivating stories from ancient philosophy, blending myth, morality, and the allure of a lost civilization.

It all begins in the dialogues "Timaeus" and "Critias," where Plato describes a magnificent empire that thrived about 9,000 years before his time. This wasn't just any empire; Atlantis was a vast and advanced civilization, located beyond the "Pillars of Hercules," what we now think may have been the Strait of Gibraltar.

According to Plato, Atlantis was a marvel of engineering and society, with a powerful navy, advanced architecture, and a complex political system. Its capital city was said to be an island of concentric rings, both of land and water, shimmering with precious metals and adorned with temples and palaces. At its heart stood a temple dedicated to the god Poseidon who, according to legend, founded Atlantis by marrying a mortal woman and siring its first kings. This was a society of abundance and knowledge, rich in resources and sophisticated in culture.

Yet, despite all its grandeur, the story of Atlantis is also a cautionary tale. Plato portrays its people as initially virtuous and just, living in harmony with the gods and with each other.

However, as time passed, they became greedy, power-hungry, and morally corrupt. Their hubris and desire to dominate other nations led them to attempt a conquest of Athens, a city Plato used to symbolize the ideal state of virtue and wisdom.

According to the philosopher, after Atlantis fell into moral decay it faced divine retribution. The gods, particularly Zeus, grew displeased with the Atlanteans' arrogance and moral corruption. In a dramatic and swift act of punishment, they decided to put an end to this once-great civilization.

The demise of Atlantis was cataclysmic. Plato describes a series of catastrophic events — earthquakes and floods of unimaginable scale — that struck the island. These natural disasters occurred in a single, terrible day and night, causing the entire island of Atlantis to sink into the depths of the ocean, disappearing beneath the waves. This sudden and complete destruction was meant to serve as a divine lesson on the consequences of arrogance and the abandonment of virtue.

With the island submerged, Atlantis was lost forever, leaving no trace of its existence. This total annihilation ensured that the story of a once great city would remain a powerful myth and serve as a moral allegory. For Plato, the tale wasn't just about the physical destruction of a city but also a warning about the dangers of losing one's moral compass. You could say that the fall of Atlantis serves as a reminder of the delicate balance between power, virtue, and the consequences that follow when that balance is disturbed.

Since Plato's time, this dramatic ending has sparked countless theories, debates, and searches for the lost city, fueling imaginations about what Atlantis might have been and why it was lost to the depths of the sea.

Ancient Egypt and Atlantis

You may be wondering, since this is a book about the wisdom of ancient Egypt, what Atlantis has to do with that. Well, the connection between the lost city of Atlantis and ancient Egypt primarily comes from the very dialogues of Plato himself.

In these works, he suggests that the story of Atlantis was preserved by Egyptian priests and later relayed to the Greeks. Here's how this connection is made:

- **Solon's Visit to Egypt:** According to Plato, the Athenian statesman Solon visited Egypt around the 6th century BCE. During his visit, Egyptian priests from the city of Sais supposedly told him the story of Atlantis. They claimed that this tale was part of their ancient records, which went back thousands of years.

- **Preservation of Ancient Knowledge:** The Egyptians were known for their extensive records and long-standing civilization. According to the priests, while other civilizations had been wiped out by floods and catastrophes, Egypt had survived due to its geography and location. Therefore, they had preserved the memory of ancient events like the rise and fall of Atlantis.

- **Atlantis and Athens:** The priests informed Solon that Atlantis had existed 9,000 years before his time and was a powerful empire that attempted to conquer the lands around the Mediterranean, including ancient Athens. This account was passed down through generations until it reached Plato, who then documented it in his dialogues.

There are also more symbolic connections. The story of Atlantis, as told by Plato, and the rich mythology of ancient Egypt both explore themes of divine punishment, the rise and fall of civilizations, and the consequences of moral decay. In Egyptian mythology, there are tales of gods punishing humanity for its transgressions, much like the gods' punishment of Atlantis.

Researchers who propose a link between Atlantis and ancient Egypt often point to a combination of historical texts, architectural similarities, and cultural parallels. While these connections are often scorned within the academic community, they provide intriguing possibilities that have fueled the search for Atlantis.

The correlation in architecture and engineering between ancient Egyptian structures and those described in Plato's account of Atlantis suggest a common source or influence. For instance, the massive stone constructions of the Egyptian pyramids and temples bear a resemblance to the grand, advanced architecture attributed to Atlantis, with its concentric rings and elaborate buildings made of precious metals.

Both Egyptian and Atlantean stories include accounts of cataclysmic events, specifically floods. The ancient Egyptians had their own myths about the "Zep Tepi" or "First Time," a golden age when gods lived among men, which ended in a great flood or series of natural disasters. Some researchers draw parallels between this and Plato's account of Atlantis, which was also destroyed by floods and earthquakes. The Egyptian "Hermetic Texts" also mention a time when the world was cleansed by water, suggesting a possible shared cultural memory of ancient catastrophes.

Other proponents of the Atlantis-Egypt connection refer to the concept of an ancient "Osirian Civilization" that predated the dynastic Egyptians and was more advanced. This idea is often associated with the mysterious structures in Abydos, specifically the Osirion, which some believe predates the rest of Egyptian civilization. Researchers like Andrew Collins, in books like "*Gateway to Atlantis*" (2000), propose that this ancient civilization might be a remnant or offshoot of Atlantis, contributing to the cultural and technological advancements seen in Egypt.

Theorists also suggest that the Sahara Desert might have been a fertile region with ancient civilizations before it became arid, potentially serving as a possible location for Atlantis or a related culture. Researchers have used satellite imagery to identify possible structures or formations beneath the sand, such as the Richat Structure (also known as the "Eye of the Sahara"), which some propose matches Plato's description of Atlantis' concentric rings. While still doubted by many scientists, this theory demonstrates the ongoing interest in finding geological links between Atlantis and Egypt.

Certain esoteric traditions and mystery schools, such as those associated with the Hermetic Order and Theosophy, claim that ancient Egyptian knowledge was inherited from Atlantis. These traditions suggest that the wisdom of Atlantis was passed down through secret societies and spiritual teachings, eventually influencing Egyptian religion and philosophy. This mystical link has also contributed to the modern mythology linking Atlantis and Egypt and we will look at it in more detail later on.

While Plato's account is the primary source linking Atlantis to Egypt, we are still looking for historical or archaeological evidence. Nonetheless, it has fueled centuries of fascination and speculation about the possible interactions between these two ancient worlds.

Written on the Wall

Ian Driscoll is an independent researcher and author with a particular focus on comparative mythology who delves into the Egyptian connection with Atlantis in a book he co-wrote with Matthew Kurtz, entitled "Atlantis: Egyptian Genesis" (2010). In it. Driscoll refers to the temple of Edfu in Upper Egypt, which contains unique and enigmatic mythological sagas, including a creation mythology referred to as the "Building Texts."

According to the Edfu mythology, an ancient island civilization was founded by the gods, ruled by twin deities, and destroyed by a great storm brought on by a malevolent serpent. After the destruction, a long period of darkness follows, until two deities emerge and restore the fallen world by erecting a pillar at the place of the fallen "djed" of the old Earth-god.

The similarities between the Edfu mythology and Plato's narrative of Atlantis, as well as the common themes found in creation myths from various cultures, could suggest that Atlantis should be understood as a myth of creation rather than a literal, historical event.

For the moment, it's worth noting that common themes present in worldwide creation myths, such as the Garden of Eden, Noah's flood, the Norse tree of Yggdrasil, and the Hindu island of Jambu, all demonstrate that Atlantis may not be solely a story about a lost civilization but a tale that reflects the timeless process of becoming and the eternal creation of the world.

Myths of 'Lost' Civilizations

As I mentioned above, the legend of a great civilization like Atlantis falling into decay and being subsequently punished by some divine intervention isn't unique. Many cultures across the world have similar tales to tell, often reflecting themes of hubris, moral decline, and divine retribution.

For example, in ancient Mesopotamian mythology, the story of the city of Uruk recounts how it was punished by the gods with a great flood due to the arrogance and corruption of its people. This myth, part of the "Epic of Gilgamesh," shares striking similarities with the biblical tale of Noah's Ark, suggesting a shared cultural memory of a great deluge wiping out an advanced civilization due to divine displeasure.

In Hindu mythology, the city of Dwarka, associated with the god Krishna, is said to have been a magnificent and advanced kingdom that eventually sank into the sea. The submergence of Dwarka is often attributed to divine will and serves as a reminder of the impermanence of even the greatest human achievements. Archaeological findings off the coast of modern-day India have sparked debates about the historical basis of this myth, further linking it to the motif of a lost civilization.

Similarly, Mesoamerican cultures like the Aztecs and the Maya have legends of ancient cities that were destroyed due to their inhabitants' moral failings. The Aztec legend of Aztlan, a mythical place believed to be the ancestral homeland of the Aztec people, speaks of a flourishing civilization that was lost to time. The Maya have tales of cities like Tollan and Xibalba, which were abandoned or destroyed due to their people's actions, often linked to their interactions with the gods or cosmic cycles.

Norse mythology also contains the tale of Ragnarok, a prophesied series of apocalyptic events that result in the destruction of the world, including the great city of Asgard. While not a "lost civilization" in the conventional sense, this myth encapsulates the idea of a powerful society being obliterated due to a combination of fate, conflict, and the wrath of higher powers.

The Greek myth of Helike, a city that was supposedly swallowed by the sea as punishment for offending the god Poseidon, offers another parallel. Ancient historians like Pausanias and Strabo wrote about Helike, suggesting that it was a real place that met a catastrophic end, much like Atlantis. Archaeological efforts have found evidence of a city submerged in the Gulf of Corinth, lending some credence to this ancient story.

Such tales, spread across different cultures and time periods, reflect a shared human fascination with the rise and fall of great societies. Whether they have a basis in historical events or are purely allegorical, they illustrate a universal narrative thread: the impermanence of human achievement in the face of greater cosmic or divine forces.

"The search for Atlantis is not just a quest for a lost city; it is an exploration of our own past and an inquiry into the potential of human civilization. Each expedition brings us closer to understanding who we are and where we came from."

— Robert Sarmast

3.2 The Search for Atlantis: Historical Expeditions and Modern Theories

The legend of Atlantis has captivated explorers, historians, and dreamers for centuries, inspiring numerous expeditions in the quest to find this lost civilization. While Plato's descriptions are the earliest and most detailed accounts of Atlantis, they have been interpreted in countless ways, leading to a variety of theories about the possible location of this ancient empire. The allure of discovering a highly advanced civilization that vanished without a trace has driven adventurers to explore diverse locations, from the depths of the Atlantic Ocean to the deserts of North Africa.

Expeditions into the Unknown - The Early Years

One of the earliest and most influential figures in the search for Atlantis was Ignatius Donnelly, a 19th-century American politician and writer. In his 1882 book, *"Atlantis: The Antediluvian World,"* Donnelly argued that Atlantis was a real, historical place that served as the cradle of all ancient civilizations. He proposed that it was located in the Atlantic Ocean, suggesting that remnants of Atlantean culture could be found in the architectural and cultural similarities between the Old and New Worlds. Donnelly's work sparked a wave of interest in the possibility of an Atlantean connection to ancient Egypt, the Maya, and other early societies.

Drawing on Plato's Dialogues, Donnelly attempted to synthesize mythology, archaeology, geology, and history into a coherent theory. He proposed that Atlantis was the mother culture of all ancient civilizations, including those in Egypt, Mesopotamia, Mesoamerica, and other parts of the world. For him, the knowledge, architecture, and religious practices of these civilizations had their roots in the advanced culture of Atlantis.

Donnelly believed that the Atlantic Ocean once contained a large landmass, which was destroyed in a catastrophic event — most likely a massive flood or series of earthquakes. He tied this idea to the numerous flood myths found across different cultures, including the biblical Great Flood in the story of Noah and the flood myths of the Sumerians and Greeks. The destruction of Atlantis, therefore, might have been the result of geological upheaval, similar to the sinking of landmasses such as Doggerland or the Sunda Shelf.

The researcher even pointed to similarities in language, symbols, and religious practices between ancient civilizations in both the Old and New Worlds as evidence of a common Atlantean origin. He pointed out that many ancient cultures shared sun-worshiping practices, pyramid-building, and similar mythological themes, which he believed were disseminated by Atlantean survivors after the destruction of their homeland.

In addition, Donnelly argued that Atlantis was the source of early human knowledge in areas like writing, metallurgy, agriculture, and navigation. He believed that Atlanteans had developed advanced technologies that were lost to history after the cataclysm that destroyed their continent. In particular, he suggested that the Atlanteans might have invented hieroglyphic writing systems, which influenced early scripts like Egyptian hieroglyphs and Mayan glyphs.

Did Donnelly have any hard evidence for his claims? Many of his theories were considered speculative by the scientific community, relying heavily on mythology, linguistic coincidences, and subjective interpretations of ancient texts. Nonetheless, his book, *"Atlantis: The Antediluvian World"* remains a cornerstone of modern Atlantis theories and has had a lasting impact on alternative history, esotericism, and pop culture. Perhaps his vision of Atlantis as the advanced progenitor of ancient civilizations continues to resonate with us because we still have so many questions about the idea of lost knowledge and forgotten history. Could the scientists be missing something?

Later Ideas about Atlantis

In the 20th century, the search for Atlantis took on new dimensions with advances in archaeology and oceanography. Edgar Cayce, often referred to as the "Sleeping Prophet", claimed to have psychic visions of Atlantis. Between the 1920s and the 1940s, Cayce gave several readings in which he described Atlantis as a highly advanced and technologically sophisticated civilization that existed from about 210,000 years ago to 10,000 BCE, with its final destruction occurring around 9,600 BCE.

According to Cayce, Atlantis was a vast continent located in the Atlantic Ocean, stretching from the Gulf of Mexico to the Mediterranean. He suggested that Atlantean civilization went through several phases of rise and fall, with a series of catastrophic events eventually leading to its submersion. His followers, along with other enthusiasts, launched underwater expeditions around the Bimini Islands in the 1960s and 1970s. In 1968, a series of underwater rock formations known as the Bimini Road was discovered, leading some to believe it was part of an ancient Atlantean structure. Could this be the case or is the Bimini Road a natural formation, as geologists like to point out? The mystery remains about the true cause of this unusual underwater landmark - is it manmade or a work of nature? Hopefully, one day, we will know for certain either way.

Apart from its location, Cayce described Atlantis in detail during his readings, referring to it as having extremely advanced technology, far surpassing modern civilization in many areas. This sophisticated technology was powered by a mysterious energy source he called "crystals" or "firestones" that harnessed both solar and earth forces. They were used for various purposes, including transportation, communication, and even warfare. He also mentioned airships, submarines, and advanced forms of healing that utilized vibrations and light, suggesting that Atlantean science had mastered both physical and metaphysical realms.

The Atlanteans also had a high degree of spiritual awareness, according to Cayce, at least in the earlier phases of their civilization. However, over time, the society became divided into two groups: the Children of the Law of One and the Sons of Belial. The first group, led by spiritual priests, sought to use Atlantean knowledge and technology for peaceful and harmonious purposes, aligning with spiritual laws. The second group were materialistic and sought to use Atlantean technology for power, control, and personal gain.

This conflict was claimed by Cayce to be the reason for the eventual downfall of Atlantis, with natural disasters caused by the misuse of advanced technologies around 50,000 BCE leading to the sinking of a portion of the continent. A second disaster happened around 28,000 BCE, when another catastrophic event submerged much of the remaining land. The final destruction, which Cayce dated around 10,000 BCE, completely submerged Atlantis, scattering its survivors across the globe.

As for the survivors, they migrated to various parts of the world, including Egypt, the Yucatan Peninsula, the Pyrenees, and the Americas. Cayce suggested that these survivors were responsible for founding the early advanced civilizations in these regions. His visions of Atlantis have been highly influential in modern New Age thought and esotericism, inspiring books, documentaries, and explorations in search of the lost city. Does it matter if the historical evidence has not yet surfaced to support his claims? That's a good question and one that continues to captivate the imaginations of those adhering to beliefs in alternative theories about history. One thing is for sure: Cayce's predictions have not yet been disproved and the mysteries of ancient civilizations remain largely lost in time.

Recent Research into Atlantis

Other explorers have looked towards the Mediterranean Sea as a possible site for Atlantis. The Greek archaeologist Spyridon Marinatos suggested that the island of Santorini, which experienced a massive volcanic eruption around 1600 BCE, could be linked to the Atlantis myth. Marinatos first made the connection in 1939, proposing that this eruption destroyed the Minoan civilization on

Crete and caused significant tsunamis. Some drew parallels with Plato's description of a powerful civilization being swallowed by the sea and the archaeologist's findings.

The Minoans, who lived on Crete, certainly had strong connections with the island of Santorini, and this could have been Plato's inspiration for Atlantis. Marinatos' excavations at Akrotiri on Santorini, which began in the 1960s, uncovered an incredibly advanced Minoan city preserved under volcanic ash, lending further support to his theory. Although he gained some scholarly support, critics argued that the timeline didn't align with Plato's account, which places Atlantis much further back in history. Was Platos' dating correct or could he have gotten it wrong? If so, the theory of Atlantis being the Minoan civilization seems pretty tempting, don't you think?

In recent years, advancements in technology have reinvigorated the search for Atlantis. With tools like satellite imagery and underwater mapping, researchers have explored various locations, including the coast of Spain, where submerged structures off the coast of Cadiz have been proposed as remnants of the lost city. One such theory, proposed by German physicist Rainer W. Kühne, suggests that the Doñana National Park in southwestern Spain could have been the site of Atlantis, citing satellite images that appear to show ring-like structures similar to Plato's descriptions.

Despite numerous expeditions and countless theories, the search for Atlantis remains one of history's most enduring mysteries. It continues to fascinate and inspire the imagination, with each new discovery adding to the wealth of speculation and wonder that surrounds this enigmatic lost civilization.

Where in the World is Atlantis?

Much has been written about the myth of Atlantis and its possible location, if real. The majority of modern researchers remain skeptical about the existence of Atlantis as a real, advanced civilization. They argue that the lack of direct archaeological evidence and the story's nature as a single-source narrative by Plato make it more plausible that Atlantis was a fictional or allegorical creation.

Some scholars suggest that Plato may have drawn inspiration from older myths and stories about lost civilizations, such as the Sumerian Epic of Gilgamesh and the biblical story of Noah's Ark. These common themes of divine punishment and the rise and fall of civilizations might have influenced Plato's narrative.

If Atlantis did indeed exist, some alternative theorists continue to seek its real location, pointing to places I mentioned previously, such as Santorini (Thera), Bimini in the Bahamas, and even Mauritania.

The Santorini Hypothesis in the Mediterranean

One of the most prominent and popular ideas is the Santorini hypothesis, which suggests that the island of Santorini, also known as Thera, is the real location of Atlantis. It proposes that the eruption of the Santorini volcano around 1600 BCE caused a catastrophic event that could have inspired Plato's story of Atlantis. But, what is the evidence for this claim?

1. Geological and Archaeological Evidence: The most compelling evidence for the Santorini hypothesis is the volcanic eruption itself, one of the largest in recorded history. The eruption devastated the Minoan civilization, which thrived on Crete and nearby islands, and created a massive caldera on Santorini. Archaeological excavations on Santorini, particularly at the site of Akrotiri, have uncovered an advanced and prosperous society with multi-story buildings, elaborate frescoes, and a sophisticated plumbing system. This civilization was abruptly buried under volcanic ash, preserving it much like Pompeii. The sudden destruction of this advanced society mirrors Plato's description of Atlantis being swallowed by the sea in a single day and night of misfortune.

2. Cultural and Technological Parallels: The Minoans, who inhabited Thera and Crete, were known for their advanced architecture, artistry, and maritime capabilities. Plato describes Atlantis as a wealthy and technologically advanced civilization with impressive architecture, intricate water systems, and a powerful navy. The Minoan culture displayed similar characteristics, including complex palaces with indoor plumbing, frescoes depicting naval prowess, and trade networks extending throughout the Mediterranean. The frescoes of Akrotiri show vibrant scenes of marine life, ships, and bustling port activities, suggesting a civilization that could have been seen as advanced and powerful, much like Plato's Atlantis.

3. Destruction by Water and Earthquakes: The Santorini eruption would have caused massive tsunamis and earthquakes, significantly impacting the surrounding areas, including Crete. Studies suggest that tsunamis generated by the eruption could have been as high as 150 feet, reaching far-off coastlines and causing widespread devastation. This aligns with Plato's account of Atlantis being destroyed by a combination of earthquakes and floods. The Minoan civilization, which was centered around the sea, would have been particularly vulnerable to such natural disasters, possibly leading to its rapid decline.

4. Alignment with Plato's Timeline: Plato mentions that Atlantis existed about 9,000 years before his time, which many scholars interpret as a mythological rather than a literal timeline. However, if we consider that Plato might have been referring to a more recent event with a different calendrical system or using symbolic numbers, the destruction of the Minoan civilization around 1600 BCE fits into a plausible historical context. This timeline places the eruption roughly 900 years before Plato, which could coincide with ancient oral traditions about a great cataclysm that inspired his narrative.

5. Location "Beyond the Pillars of Hercules": One challenge in linking Santorini to Atlantis is Plato's reference to Atlantis being "beyond the Pillars of Hercules" (modern-day Strait of Gibraltar). However, some scholars argue that this phrase could be metaphorical or that "beyond" might simply mean "in the western Mediterranean." Given the Minoans' extensive trade routes and influence throughout the Mediterranean, it's possible that the story of their destruction could have traveled westward and evolved into the legend of Atlantis.

In *"The Story of Atlantis and Its Link to the Santorini Eruption: A Geomythological Perspective"* (2014), S. Lucrezi & M. Saayman. explore the geological and cultural aspects of the Santorini eruption in relation to the Atlantis myth. W. L. Friedrich also provides an in-depth analysis of the Santorini eruption in his work, *"Santorini: Volcano, Natural History, Mythology"* (2009). The author talks in his book of the impact of the eruption on the Minoan civilization, and its potential link to the Atlantis narrative.

While the Santorini hypothesis remains one of the most popular theories, it's important to note that it is still speculative. The idea that Santorini is Atlantis hinges on the interpretation of Plato's account as a blend of historical events and allegory. Despite the lack of direct evidence confirming Santorini as Atlantis, the parallels between the catastrophic volcanic eruption and the sudden destruction of an advanced society provide a compelling argument that continues to intrigue researchers and enthusiasts alike.

The Azores in the Atlantic Ocean

Some researchers propose that the Azores archipelago, situated in the mid-Atlantic, could be the remnant of Atlantis. According to this theory, a substantial landmass once existed in this region and sank due to tectonic activity, which aligns with Plato's description of Atlantis's catastrophic end.

1. Geological Activity: The Azores are located on the mid-Atlantic Ridge, a major tectonic boundary where the Eurasian and North American plates meet. This region is characterized by significant geological activity, including earthquakes and volcanic eruptions. Proponents of the Azores theory argue that such tectonic forces could have led to the subsidence of a large landmass, similar to the description of Atlantis sinking into the ocean due to natural disasters.

2. Geological Evidence: Research into the tectonic activity of the mid-Atlantic Ridge supports the possibility of significant land movements. The region experiences frequent volcanic activity and seismic events, which could have contributed to the gradual submergence of land.

3. Submarine Structures: Submarine exploration around the Azores has revealed various underwater formations that some researchers believe could be remnants of ancient man-made structures. These include linear stone formations, terraced platforms, and other geometric patterns observed on the ocean floor. The discovery of underwater formations near the Azores, such as what some describe as "ancient roads" and "stone walls," has led to speculation that these could be remnants of an advanced civilization. While some argue these formations resemble man-made structures, others contend they are natural geological features.

In *"The Lost Continent: A Study of the Evidence"* (1989), David Rupp discusses the geological and archaeological findings in the Azores and their potential connection to the Atlantis legend. Supporters of the Azores hypothesis point to ancient maps and texts that reference a large landmass in the Atlantic, suggesting it could correspond to the location of Atlantis. They argue that such references, combined with the geological evidence, support the idea of a lost civilization in this region. Researchers have created geological models that simulate the effects of tectonic activity on the mid-Atlantic Ridge, demonstrating how a large landmass could have sunk due to shifts in the Earth's crust. These models help illustrate how such a phenomenon could align with the Atlantis narrative.

Graham Hancock discusses the possibility of lost civilizations in the Atlantic and explores the evidence supporting the Azores theory in his book *"Underworld: The Mysterious Origins of Civilization"* (2002). Equally, Robert Ballard talks about explorations of various underwater formations and their potential connections to the Atlantis legend, including references to the Azores, in his work *"The Discovery of the Lost Continent"* (2010).

The theory that the Azores could be the remnants of Atlantis is supported by geological activity and certain underwater formations that might suggest human intervention. However, while the evidence presents intriguing possibilities, there is no definitive proof linking the Azores to the lost city of Atlantis as described by Plato. As with other theories, the connection remains open to interpretation, reflecting the ongoing quest to solve one of history's most enduring mysteries.

The Richat Structure (Eye of the Sahara) in Mauritania

The Richat Structure, often referred to as the "Eye of the Sahara," is located in the Sahara Desert of Mauritania. It has been proposed as the site of Atlantis due to its striking circular and ring-like appearance. This structure is thought by some researchers to match Plato's description of Atlantis's capital city, which he described as having concentric rings of land and water.

1. Geometric Appearance: The Richat Structure is a massive circular formation, about 40 kilometers (25 miles) in diameter, consisting of concentric rings of exposed rock. Its layout, with a central area surrounded by rings of land and water, closely resembles Plato's depiction of Atlantis, which described a city with a central island encircled by alternating rings of land and water. The concentric rings of the Richat Structure appear to mirror the geographical layout described by Plato, which includes a central island with rings of land and water radiating outward. This resemblance has led some to suggest that it could be the remnants of the lost city.

2. Size and Scale: The sheer size and scale of the Richat Structure support the idea that it could have accommodated a large, advanced civilization. Its dimensions are large enough to have potentially housed an extensive urban area and infrastructure, aligning with Plato's description of a grand and complex city. The structure's size and the visible features of its concentric rings have prompted researchers to speculate that it could have been a significant urban center, possibly linked to the Atlantis narrative.

3. Geological and Historical Considerations: Some researchers argue that the Richat Structure's formation may have resulted from ancient tectonic activity and erosion, which could have created the circular pattern over millennia. This natural formation could potentially be mistaken for the remnants of an advanced civilization. Studies suggest that the structure is a dome created by erosion of the surrounding rock layers. While this geological explanation accounts for its circular appearance, it does not preclude the possibility that ancient people might have interpreted it as a site of historical or mythological significance.

In *"Atlantis in the Sahara: The Ancient World's Most Resilient Myth Unveiled?"* (2004), George Alexander and Natalia Koutik explore the possibility that ancient civilizations might have existed in the Sahara region and examine the Richat Structure as a potential site for Atlantis. Humberto Cruz also discusses the Richat Structure's alignment with Plato's description of Atlantis and the geological evidence supporting its natural formation in *"The Eye of the Sahara: The Lost City of Atlantis?"* (2008)

The "Eye of the Sahara" presents an intriguing case for being the location of Atlantis due to its concentric ring-like appearance, which mirrors Plato's description of the city. However, as with other theories, definitive evidence linking the Richat Structure to the historical Atlantis remains elusive, and the connection remains speculative and open to interpretation.

The Bimini Island Theory in the Bahamas

One of the intriguing theories about the location of Atlantis places it near Bimini Island in the Bahamas. This idea gained popularity after the discovery of a peculiar underwater formation in 1968, known as the Bimini Road. Proponents of this theory suggest that these submerged structures are remnants of the lost civilization of Atlantis.

1. Description: The Bimini Road is a series of large, flat, rectangular limestone blocks located off the coast of North Bimini in the Bahamas. The stones appear to be arranged in a linear or curving path, resembling a road or wall. The formation spans about half a mile (0.8 km) in length and is situated in relatively shallow water, making it accessible for exploration.

2. Man-Made Structures: Advocates of the Atlantis-Bimini theory argue that the arrangement and shape of the stones suggest they are man-made, rather than a natural geological formation. Some believe these could be remnants of walls, roads, or other structures from a once-thriving civilization. The precision and regularity of the blocks lead to speculation that they were part of a harbor or an ancient city that was submerged due to rising sea levels or a catastrophic event.

3. Connection to Edgar Cayce's Prophecy: Interest in Bimini as a possible location for Atlantis was fueled by the American psychic Edgar Cayce, who in the 1930s predicted that evidence of Atlantis would be found near Bimini in the late 1960s. When the Bimini Road was discovered in 1968, it was seen by some as a fulfillment of this prophecy. Cayce's followers believe that Bimini was part of a network of Atlantean islands and that the structures found there are remnants of this ancient culture.

4. Unusual Geology: Supporters of the theory argue that the stones of the Bimini Road resemble cut and placed blocks, much like those used in ancient construction. The linear and rectangular shapes are seen as unnatural for typical beachrock formations. Some researchers have even proposed that the stones show signs of being worked or tooled by human hands.

Most mainstream geologists and archaeologists believe that the Bimini Road is a natural formation known as beachrock, which forms in coastal areas through the natural cementation of sedimentary rock. Beachrock can fracture and create regular patterns that resemble man-made structures. The flat, rectangular shapes and arrangement of the stones can be explained by natural processes, such as the fracturing and shifting of the rock over time due to tidal movements and weathering.

Studies of the Bimini Road stones also suggest they date back to around 2,000 to 3,000 years ago, which is much later than the supposed time of Atlantis's existence as per Plato's account (around 9,000 years before his time, or roughly 11,600 years ago). This discrepancy in dating weakens the theory that Bimini is Atlantis, although the allure of this underwater formation continues to inspire explorations and theories surrounding the lost city.

Antarctica

One speculative theory posits that Atlantis lies beneath the ice of Antarctica. Proponents suggest that the continent was once situated in a more temperate region before shifting to its current icy position due to crustal displacement or a polar shift. This theory aims to align with Plato's description of Atlantis as a once-great civilization that vanished due to a cataclysmic event.

1. The Piri Reis Map: The Piri Reis Map, a 16th-century cartographic artifact, is often cited by proponents of this theory. The map appears to depict the coastline of Antarctica without its ice cover, suggesting that ancient cartographers may have had access to knowledge about the continent's geography before it was known to modern explorers. Some researchers argue that the map's accurate depiction of the Antarctic coastline could indicate that it was drawn from earlier sources, possibly referencing a time when Antarctica was ice-free and habitable. This has led to speculations that advanced civilizations, such as Atlantis, might have existed there. In his book, *"Maps of the Ancient Sea Kings: Evidence of Advanced Civilization in the Ice Age"* (1966), Charles Hapgood proposed that maps like the Piri Reis could be evidence of ancient civilizations with knowledge of Antarctica, suggesting that the continent's earlier, ice-free state could correspond to Plato's Atlantis.

2. Geological Evidence: Studies show that Antarctica was once a much warmer, ice-free continent, with a climate that could have supported vegetation and potentially advanced human settlements. Fossil evidence of subtropical plants and ancient coal deposits indicates that Antarctica had a temperate climate in the distant past. Geological studies and paleoclimatic models support the idea that Antarctica was once positioned closer to the equator and had a suitable environment for advanced life before tectonic shifts moved it to its current icy position.

3. Crustal Displacement and Polar Shift: Proponents argue that a crustal displacement or pole shift could explain the dramatic change in Antarctica's climate and geography. According to this theory, such shifts could have led to the submergence of advanced civilizations located on the continent. Researchers have proposed various models of polar shifts and tectonic movements that could explain the dramatic changes in Antarctica's position and climate over millions of years.

William Wegman explores the concept of polar shifts and their potential effects on global climates, including the implications for historical civilizations in his book, "*The Theory of Polar Shifts and Their Impact on Earth's Climate*" (2003).

Southern Spain – The Doñana National Park

It is believed by some researchers that Atlantis was located in what is now southern Spain, near Doñana National Park. This theory proposes that the area was part of a large, complex society that was eventually submerged by catastrophic flooding and tsunamis.

1. Submerged Structures: Archaeological surveys and underwater explorations in the Doñana National Park region have uncovered submerged structures that some researchers interpret as remnants of an ancient civilization. These findings include large circular formations and other features that resemble the concentric rings described by Plato. In *"Finding Atlantis: A Scientific Expedition to the Sunken City"* (2011), Richard Freund features a detailed study of the Doñana region, using satellite imagery, underwater exploration, and ground-penetrating radar to support the theory that this area could be the site of Atlantis.

2. Geological and Environmental Conditions: The Doñana region is prone to seismic activity and flooding, which aligns with Plato's account of Atlantis's destruction. The area's susceptibility to tsunamis and natural disasters supports the idea that a large landmass could have been submerged due to such catastrophic events. In another of his works, entitled *"The Lost City of Atlantis: The Evidence from Southern Spain"* (2012), Freund provides an overview of the geological and archaeological evidence supporting the theory that the Doñana region could be linked to the Atlantis legend.

Both the Antarctica and Southern Spain theories offer intriguing possibilities for the location of Atlantis, based on geological, cartographic, and archaeological evidence.

While the Piri Reis Map and the geological evidence from Antarctica suggest ancient knowledge of a now-submerged landmass, the discoveries in Doñana National Park propose that southern Spain could be the site of the lost city. However, no conclusive evidence has yet been found to definitively link these locations to the Atlantis described by Plato.

Considering the various theories about the possible location of Atlantis, there is no consensus among researchers, and definitive evidence remains elusive. While archaeological discoveries and geological evidence in places like Santorini and southern Spain have sparked debate and speculation, they have not conclusively proven the existence of Atlantis as described by Plato. As such, Atlantis continues to straddle the line between myth and history, captivating the imagination of those who seek to uncover its mysteries.

"The remnants of the Atlantean civilization whisper through the ages, a reminder of an advanced society that fell into ruin, leaving behind cryptic knowledge and strange legends that continue to haunt our dreams."

— H.P. Lovecraft

3.3 Atlantean Technology and Civilization: Legends and Speculations

What really lies behind the myth of Atlantis? Are we any nearer to discovering the truth about this legendary land or does scientific research only prevent us from being open to other explanations?

While archaeologists, geologists, and other scientists like to have hard evidence in their hands, there are many other voices who have researched the myth of Atlantis with a more open-minded approach. Different theories have been put forward as to the meaning behind the myth, as well as possible interpretations of it. Let's take a look at some of them below:

Atlantis as a Lost Advanced Civilization

Some theories propose that Atlantis was a highly advanced civilization possessing technology and knowledge far beyond what was known in the ancient world, which was eventually lost or destroyed. These theories often delve into the realms of speculative history, suggesting that Atlantis was either an indigenous advanced culture that developed extraordinary technologies or a civilization that had contact with or was influenced by ancient aliens. While these ideas are controversial and not widely accepted in mainstream archaeology, they continue to fascinate and intrigue us, partly due to the enigmatic nature of ancient technological achievements that we still struggle to fully understand.

1. Technological and Architectural Marvels: Proponents of the advanced civilization theory often point to the sophisticated engineering and architectural feats of ancient cultures as evidence that Atlantis could have possessed advanced technology. For instance, the precision with which the Great Pyramid of Giza was constructed, the complex stonework at sites like Machu Picchu and Ba'albek, and the alignment of these structures with celestial bodies have led some to speculate that these ancient sites might have been influenced by a civilization like Atlantis. They argue that the Atlanteans could have possessed advanced tools and knowledge, such as anti-gravity technology, energy manipulation, or even a form of electricity, which enabled them to construct such monumental works.

2. Myths and Legends of Advanced Beings: Many ancient cultures have myths and legends about gods or beings from the sky who brought advanced knowledge to humanity. In ancient Sumerian texts, the Anunnaki are described as deities who imparted wisdom and technology to humans. Similarly, the Egyptian god Thoth and the Greek Titan Prometheus are associated with the dissemination of knowledge and technology. Some theorists suggest that these myths could be distorted memories of the Atlanteans or other advanced beings who interacted with early human societies, teaching them advanced techniques in agriculture, architecture, and astronomy.

3. Advanced Maritime Capabilities: Plato describes Atlantis as a powerful naval empire with a fleet capable of navigating the world's oceans. This idea aligns with theories that ancient civilizations might have had advanced maritime capabilities, enabling them to explore and colonize distant lands. Researchers like Thor Heyerdahl, who famously sailed the Kon-Tiki to demonstrate the possibility of transoceanic contact between ancient cultures, argue that the exchange of knowledge and technology could have occurred between Atlantis and other civilizations. The Piri Reis Map has also been cited as evidence that ancient civilizations had advanced cartographic and navigational skills, potentially inherited from Atlantis.

4. Evidence of Ancient High Technology: There are numerous artifacts and archaeological findings that suggest the existence of ancient high technology. For example, the Antikythera Mechanism, an ancient Greek analog computer used to predict astronomical positions, indicates a level of technological sophistication previously thought impossible for that time. Similarly, the Baghdad Battery is believed by some to be an ancient galvanic cell. These artifacts could be remnants or descendants of Atlantean technology, representing just a fraction of what this advanced civilization might have achieved.

5. Theories of Ancient Aliens and Extra-terrestrial Influence: A more fringe branch of the Atlantis theory involves the idea of ancient aliens, suggesting that the technological prowess of Atlantis was due to contact with extraterrestrial beings. Proponents of this theory argue that ancient texts and artwork from various cultures depict flying machines, advanced machinery, and beings with otherworldly appearances. They posit that these representations are not merely symbolic but are instead historical records of interactions with advanced alien visitors who may have established colonies on Earth, with Atlantis being one of the most advanced.

In his book, "*Technology of the Gods: The Incredible Sciences of the Ancients*" (2000), David Childress explores the idea that ancient civilizations, including Atlantis, possessed advanced technologies that have since been lost or suppressed. Erich von Däniken also alluded to the idea that ancient civilizations were influenced by extraterrestrial beings, which could include Atlantis, in his book "*Chariots of the Gods? Unsolved Mysteries of the Past*" (1968). In "*Fingerprints of the Gods: The Evidence of Earth's Lost Civilization*" (1995), Graham Hancock Investigates various ancient sites and artifacts, suggesting that a highly advanced civilization did indeed exist and was lost to history, potentially influencing known ancient cultures.

While mainstream archaeology tends to attribute the technological achievements of ancient civilizations to human ingenuity and gradual development over time, the theory of Atlantis as a lost advanced civilization presents an alternative narrative. It challenges our conventional understanding of history by suggesting that there was once a golden age of knowledge and technology that has been obscured by the mists of time. Whether or not these theories hold any truth, they keep the legend of Atlantis alive and continue to inspire curiosity about humanity's past and the possibility of forgotten epochs of advanced civilization.

Atlantis as a Mythological Allegory

Many scholars interpret Atlantis as a mythological allegory created by Plato to convey philosophical or moral lessons. According to this perspective, Atlantis was never meant to be a historical account but rather a fictional narrative designed to illustrate key ideas about human nature, society, and the consequences of moral decay. In this interpretation, Plato uses the story of Atlantis as a literary device to explore themes of hubris, the rise and fall of civilizations, and the tension between the ideal and the real.

1. Plato's Philosophical Intentions: Plato often used allegory and myth in his works to explore complex philosophical ideas. The most famous example is the Allegory of the Cave in "*The Republic*," where he describes prisoners in a cave mistaking shadows for reality, illustrating his ideas about ignorance, enlightenment, and the nature of reality. Similarly, in the dialogues of "*Timaeus*" and "*Critias*," where Atlantis is mentioned, Plato presents the story as a means "to discuss broader philosophical themes, such as the dangers of pride and the importance of virtuous governance. Atlantis, in this context, may serve as an allegory for the ideal society that falls from grace due to moral corruption and overreach, echoing the philosophical concerns that Plato frequently explored.

2. The Moral of Hubris and Divine Retribution: Atlantis's narrative follows a pattern common in Greek mythology, where a once-great civilization falls due to excessive pride and arrogance that leads to divine punishment. This theme can be seen in other Greek myths, such as the story of Icarus, who flies too close to the sun, or Oedipus, whose attempt to outsmart fate leads to his downfall. By constructing Atlantis as an allegorical tale, Plato might have been cautioning against the dangers of conceit, both at the individual and societal levels. The story serves as a reminder that even the most advanced and powerful civilizations are not immune to moral decay and that such decay can lead to catastrophic consequences.

3. Ideal Society and the Human Condition: Scholars argue that Atlantis represents Plato's vision of an ideal society that eventually succumbs to human weaknesses. In Critias, he describes Atlantis as a prosperous, technologically advanced society governed by a just and wise leadership. However, as the Atlanteans grow more powerful, they become greedy, corrupt, and seek to conquer other lands, ultimately leading to their downfall. This narrative mirrors Plato's ideas in *The Republic*, where he outlines his vision of an ideal state governed by philosopher-kings who are wise and virtuous. The fall of Atlantis can be seen as a metaphor for the fragility of utopias and the inherent flaws in human nature that make it difficult to maintain an ideal society.

4. Historical Context and Athenian Ideals: Plato sets the Atlantis narrative in a mythical past where Athens stands as a bastion of virtue and justice. This portrayal reflects Plato's admiration for Athenian ideals and the city-state's values of democracy, military prowess, and intellectual achievement. By contrasting Atlantis's moral decline with the virtuous stand of ancient Athens, Plato underscores the moral superiority of his own city-state, using the story to celebrate Athenian resilience and ethical governance. It serves as an exhortation to his contemporaries to uphold these values and avoid the pitfalls of excess and corruption.

5. Critiques of Contemporary Societies: Plato might have used the allegory of Atlantis to critique contemporary societies, particularly the imperial ambitions and moral decline of his own Athens during the Peloponnesian War. The parallels between Atlantis's expansionist desires and the Athenian empire's imperialism suggest that Plato was using the story to reflect on the dangers of militaristic expansion and the loss of moral integrity. Atlantis, in this sense, becomes a cautionary tale about the potential consequences of a society that loses sight of justice and virtue in pursuit of power and wealth.

By interpreting Atlantis as a mythological allegory, scholars suggest that Plato crafted a narrative that goes beyond mere storytelling to provide a profound commentary on the human condition. The tale of Atlantis serves as a mirror reflecting the potential and pitfalls of civilization, challenging readers to consider the ethical and moral dimensions of societal development. It remains a powerful example of how myth can be employed to explore and communicate complex philosophical ideas, illustrating the timelessness and depth of Plato's work.

Atlantis as a Dramatic Representation of Quantum Mechanics

This fringe theory suggests that Plato's story of Atlantis is not a historical account but rather a metaphorical representation of quantum mechanics, with Atlantis symbolizing the hidden world of quantum reality. Proponents of this theory argue that Plato, known for his deep philosophical inquiries into the nature of reality, used the Atlantis narrative as an allegory for the complex and often counterintuitive principles found in quantum mechanics. While this idea falls outside the mainstream interpretation of Plato's work, it opens an intriguing dialogue between ancient philosophy and modern physics.

1. Plato's Dualism and the Nature of Reality: Plato's philosophy often dealt with the concept of dualism — the division between the physical world and the world of forms or ideals. In the context of quantum mechanics, this dualism can be likened to the difference between the observable, macroscopic world and the underlying quantum reality that governs it. Atlantis, in this theory, represents the unseen quantum world that exists alongside our perceivable reality. Just as Atlantis was said to have existed in a different realm and was lost to the everyday world, the quantum realm is hidden from direct observation, only revealing its effects through phenomena like wave-particle duality and quantum entanglement.

2. Metaphorical Interpretation of Destruction: The sudden disappearance of Atlantis can be seen as a metaphor for the elusive nature of quantum particles, which can appear and disappear or change states when observed. The dramatic destruction of Atlantis by a cataclysm could symbolize the collapse of a quantum wave function — the process by which a quantum system transitions from a superposition of states to a single state upon observation. This interpretation draws a parallel between the unpredictable and transformative nature of quantum events and the sudden loss of an advanced civilization like Atlantis.

3. The Concept of Potentiality: In quantum mechanics, particles exist in a state of potentiality, where they have the potential to be in multiple states at once until they are observed. This concept can be metaphorically linked to the idea of Atlantis, which, while not directly observable or verifiable, exists as a potential within the realm of human consciousness and mythology. The story of Atlantis, therefore, could be seen as Plato's attempt to explore the nature of potential realities and how they can influence the physical world, much like quantum potentiality affects outcomes in the observable universe.

4. Atlantis as a Hidden Dimension: Some interpretations of quantum mechanics involve the idea of multiple dimensions or parallel universes, where different versions of reality coexist. In this context, Atlantis could symbolize a hidden or parallel dimension that once intersected with our own. This idea aligns with certain quantum theories like the "many-worlds interpretation," which suggests that every possible outcome of a quantum event actually occurs in a separate, parallel universe. Atlantis, as a lost and hidden world, could represent one such dimension that once interacted with our reality but has since diverged or become inaccessible.

5. The Role of the Observer: Quantum mechanics places significant importance on the role of the observer in determining the state of a quantum system. Similarly, the story of Atlantis only exists because it was recorded and shared by observers like Plato. The theory suggests that Atlantis, as a metaphor, highlights the relationship between the observer and the observed, echoing the quantum principle that the act of observation affects the reality being observed. In this sense, Atlantis serves as a narrative device to explore how our perception shapes the reality we experience.

In *The Tao of Physics* by Fritjof Capra (1975), we read about the parallels between modern physics and ancient mysticism, including the philosophical implications of quantum mechanics that align with metaphysical interpretations. More recently, in *"Quantum Reality, the Emperor's New Mind, and Atlantis"* (2022) by C. C. King, we discover a work that delves into the relationship between quantum mechanics and ancient philosophical ideas, including a speculative look at Atlantis as a metaphorical representation of quantum principles.

While the theory that Atlantis is a representation of quantum mechanics is more philosophical than scientific, it highlights the enduring nature of Plato's work as a source of inspiration for exploring complex ideas about reality. By interpreting Atlantis through the lens of quantum mechanics, we receive a unique perspective that bridges ancient wisdom with the cutting-edge theories of modern physics.

Atlantis as a Symbol of a Golden Age

The concept of Atlantis as a representation of a "Golden Age" is a significant and enduring interpretation of Plato's myth. This perspective suggests that Atlantis was not just a real place but also a symbol of an idealized period of prosperity, wisdom, and harmony that contrasts sharply with the more flawed societies of Plato's time. Here's an in-depth exploration of this idea:

1. Plato's Description of Atlantis: In his dialogues "*Timaeus*" and "*Critias*", Plato describes Atlantis as a powerful and technologically advanced civilization that existed around 9,000 years before his own time. It was a magnificent empire with impressive architecture, sophisticated technology, and a well-organized society, characterized by its prosperity, opulence, and moral virtues before its eventual fall into corruption and destruction. Plato's description of Atlantis as a utopian society with an ideal social structure, was a place where the ruling class was composed of wise and virtuous leaders. This idealization aligns with the notion of a golden age — a period of unmatched harmony and excellence.

2. Atlantis and the Myth of the Golden Age: The concept of a golden age is a recurring theme in many mythologies and philosophies, often symbolizing a time of peace, prosperity, and enlightenment. Atlantis is frequently linked to this archetype, representing a lost era of perfection that serves as a contrast to contemporary times. Similar to the Atlantis narrative, many cultures have myths about a lost golden age or paradise. For example, the Garden of Eden in the Judeo-Christian tradition represents an idyllic state of innocence and harmony before the fall of humanity. The Hindu concept of Satya Yuga, or the "Golden Age," also describes an era of truth and righteousness that precedes the current age of moral decline.

3. Cultural and Historical Reflections: The idea of Atlantis as a golden age reflects a cultural and historical longing for a more perfect past. In periods of social or political upheaval, the myth serves as a nostalgic reference to an idealized past that contrasts with contemporary challenges. Plato's own society, Athens, was undergoing significant changes and faced political and moral issues at the time. The Atlantis myth could be seen as a reflection of his desire for a return to an idealized state of civilization.

What we can say is that, through Plato's account, Atlantis serves as both a narrative of a once-great civilization and a philosophical allegory about the virtues and vices that shape societies. This interpretation emphasizes the myth's enduring relevance as a reflection of humanity's longing for a perfect past and the lessons to be learned from its fall.

Atlantis, Theosophy, and Other Spiritual Movements

Helena Petrovna Blavatsky, a 19th-century Russian mystic and co-founder of the Theosophical Society, introduced a unique and esoteric perspective on Atlantis. Her interpretation of the lost city was deeply rooted in Theosophy, a spiritual movement blending Eastern and Western religious philosophies, science, and mysticism. Her ideas had a great impact on esotericism in the West, with her theory about Atlantis being a home of a spiritually advanced race being quite controversial by contemporary standards.

In her magnum opus, "*The Secret Doctrine*" (1888), Blavatsky presents Atlantis as the home of the "Fourth Root Race" — a spiritually advanced race that existed millions of years ago. According to her, humanity is divided into "Root Races," each representing different stages of spiritual and physical evolution.

The Atlanteans, according to Blavatsky, were the Fourth Root Race, possessing advanced knowledge and psychic abilities. They were considered superior in intellect and spirituality to modern humans and were capable of performing what would now be considered supernatural feats.

Blavatsky attributed advanced technology and spiritual practices to the Atlanteans, including flying machines (similar to the "*Vimanas*" mentioned in Hindu texts) and the ability to harness natural energies. She claimed they had profound knowledge of the natural world, which they used to create a technologically advanced society.

The fall of Atlantis, according to Blavatsky, was a result of the Atlanteans' moral and spiritual decline. She suggested that the ancient civilization misused its powers and advanced technology for selfish and materialistic purposes, bringing about its own destruction through a series of natural disasters, including earthquakes and floods. The narrative aligns with the idea of divine retribution seen in other mythologies and reinforces the Theosophical concept of karma and cosmic justice.

For the Theosophists, the destruction of Atlantis served as a turning point in the spiritual evolution of humanity. The lessons from the fall of Atlantis were integral to the development of the "Fifth Root Race" (the current humanity), which must learn from the mistakes of the past to achieve spiritual enlightenment.

Blavatsky's interpretation of Atlantis had a significant impact on esoteric and occult traditions in the late 19th and early 20th centuries. Her ideas were incorporated into various strands of spiritual thought, including Anthroposophy, a spiritual philosophy founded by Rudolf Steiner. The thinking behind this school of thought is to integrate science, art, and spirituality, with one of Steiner's more controversial ideas involving his interpretation of the mythical Atlantis. he incorporates Atlantis into his broader worldview by presenting it as a real ancient civilization that was pivotal in the spiritual evolution of humanity.

In his lectures, Steiner describes Atlantis as an advanced civilization that existed between modern Europe and America. According to him, Atlanteans possessed unique spiritual abilities that far surpassed those of modern humans. They had capacities for telepathy, spiritual insight, and a deeper connection to the cosmos. Over time, these spiritual faculties declined, leading to the eventual fall of Atlantis.

Steiner's Anthroposophy places human evolution in a spiritual context, describing how humanity has evolved through different epochs, each with its own distinctive soul-development. He taught that Atlantis was a key stage in this evolutionary journey, where humanity was transitioning from a purely spiritual consciousness to a more material-based consciousness.

Ultimately, according to Steiner, Atlantis was destroyed in a series of natural disasters, primarily due to misuse of their spiritual and technological knowledge. He suggested that the fall of this advanced civilization was not just a physical event but also symbolized a shift in human consciousness, where materialism began to dominate over spiritual wisdom.

This account serves as another moral lesson for modern society, since humanity is at risk of making the same mistakes as the Atlanteans by overemphasizing materialism and technology while neglecting spiritual development. Steiner believed that Anthroposophy, with its emphasis on harmonizing spiritual and scientific knowledge, could help guide humanity away from repeating Atlantis' errors.

The New Age movement of the 20th century also adopted Blavatsky's views on Atlantis, incorporating the idea of a lost, spiritually advanced civilization into its narrative of human evolution and enlightenment. Blavatsky's ideas also influenced a wide range of literature and fiction, inspiring stories about lost civilizations, ancient wisdom, and the spiritual potential of humanity.

Such theories about Atlantis have been widely criticized for their lack of empirical evidence and reliance on esoteric knowledge. Opponents argue that the accounts put forward by Blavatsky were speculative and lacked any grounding in historical or archaeological facts. Equally, some of her ideas have been slammed for promoting pseudoscience and for containing elements of racial theory that are considered problematic by modern day standards. Her concept of "Root Races" in particular has been criticized for its hierarchical and Eurocentric implications.

Nonetheless, Blavatsky's blending of spiritual evolution, advanced technology, and moral decline created a mythos that resonated with esoteric and occult traditions, continuing to influence modern interpretations of Atlantis within the context of spiritual and metaphysical thought.

The Hermetic tradition, rooted in the teachings attributed to Hermes Trismegistus, a syncretic figure combining the Greek god Hermes and the Egyptian god Thoth, is rich in mystical and philosophical ideas. These teachings were highly influential in Hellenistic Egypt and later Western esoteric traditions.

Hermeticism and Egypt

The Hermetic texts, also known as the *Hermetica*, are a collection of writings attributed to Hermes Trismegistus, a legendary figure who is a fusion of the Greek god Hermes and the Egyptian god Thoth. These texts form the basis of *Hermeticism*, a spiritual and philosophical tradition that emerged in the Greco-Roman world, particularly in Hellenistic Egypt. The Hermetic texts encompass a variety of subjects, including theology, cosmology, philosophy, astrology, alchemy, and the nature of the divine.

They were primarily written between the 1st and 4th centuries CE, during the Hellenistic period in Egypt. After the conquests of Alexander the Great, Greek culture came into close contact with ancient Egyptian religious traditions. It was this fusion of Greek philosophy and Egyptian spirituality that played a significant role in the creation of Hermeticism.

While not overtly focused on Atlantis in its earliest texts, Hermeticism is often linked to Atlantean lore by later esoteric traditions. The connection emerges more strongly in later occult interpretations that seek to integrate Hermetic and Atlantean wisdom into a broader narrative of human spiritual evolution.

According to some esoteric thinkers, Thoth (or Hermes Trismegistus) was believed to have been an Atlantean priest-king who carried the wisdom of Atlantis to Egypt. This idea posits that after the fall of Atlantis, Thoth fled to Egypt, bringing with him the knowledge of the ancient Atlanteans, which later formed the foundation of Egyptian and Hermetic wisdom traditions.

One of the most famous Hermetic texts, the *Emerald Tablet,* attributed to Hermes Trismegistus, is said to encapsulate ancient, timeless wisdom. Occultists and esoteric thinkers have often speculated that the knowledge in this tablet could originate from the lost civilization of Atlantis. The Tablet's cryptic phrase "As above, so below" reflects Hermeticism's emphasis on the interconnectedness of the macrocosm and microcosm, which some have linked to Atlantean teachings about the unity of the material and spiritual worlds.

Blavatsky, the founder of Theosophy, suggested that the spiritual teachings preserved in Hermeticism were part of a broader ancient wisdom tradition that included both Atlantean and Lemurian influences. She claimed that Egypt, as the cradle of post-Atlantean civilization, inherited Atlantean knowledge, which later formed the basis of its mystery schools and Hermetic teachings.

The Hermetic tradition, especially as it developed in later esoteric schools, emphasized initiation into ancient knowledge through mystery schools, secretive societies that taught sacred wisdom to chosen students. According to certain occult perspectives, these mystery schools were continuations of Atlantean spiritual teachings, brought to Egypt by survivors of Atlantis and preserved by the priesthood.

Many Hermetic thinkers interpret Egyptian symbols — such as the pyramids, the Eye of Horus, and the Sphinx — as containing hidden wisdom that links back to Atlantis. These symbols are seen as containing spiritual truths about the cosmos, human nature, and the process of enlightenment, a process that, according to some, was mastered by the Atlanteans and then transmitted to the Egyptians.

While Hermetic texts do not explicitly recount the fall of Atlantis, later esoteric traditions often build on Hermetic cosmology to interpret Atlantis' destruction as part of a larger spiritual and cosmological narrative. Echoing Steiner's ideas, many esotericists suggest that the downfall of Atlantis was due to the Atlanteans' misuse of their spiritual knowledge. Hermeticism's emphasis on personal enlightenment and the pursuit of divine wisdom is often viewed as an attempt to preserve the spiritual teachings that the Atlanteans had mastered before their fall.

Hermeticism describes the cosmos as a dynamic interplay between divine forces and the material world. Esoteric interpreters of Hermeticism often draw parallels between the decline of Atlantis and Hermetic cosmology, seeing the material world as a place where spiritual beings are at risk of becoming ensnared by physical desires. This echoes the idea that the Atlanteans lost their way by becoming too materialistic and disconnected from their divine nature.

Is there any evidence to support the view that Egypt, with its emphasis on spiritual wisdom, inner transformation, and the pursuit of divine knowledge, could be linked to the myth of Atlantis? Was Egypt really a repository of Atlantean wisdom, with Hermes Trismegistus (Thoth) serving as a key figure who transmitted Atlantean spiritual teachings to Egypt? A significant part of the Western esoteric tradition believes so but what are your views on this possibility?

The Saga of the Sunken Cities

While the idea of an ancient sunken city lying at the bottom of the ocean that once belonged to an advanced civilization fuels our imagination, there are many sites to choose from. Which one could be Atlantis?

At present, there are over 200 known sunken cities around the world, submerged due to natural disasters, rising sea levels, and other catastrophic events. These underwater cities have been discovered in various regions, from the Mediterranean to the coasts of Asia and the Americas. Each offers unique insights into ancient civilizations and their fates, with many still being explored by archaeologists and marine scientists.

In relation to ancient cities, many still hold onto their secrets in their subterranean resting places. Here are some examples of ancient sunken cities around the world, each offering a glimpse into civilizations that thrived before being submerged due to natural events or human activities:

- ◆ **Heracleion (Thonis) – Egypt**

    Founded around the 8th century BCE near the Nile Delta, Thonis sank around the 2nd century CE. The city was submerged due to earthquakes and soil liquefaction, with remnants of temples, colossal statues, treasures, and shipwrecks still visible underwater.

- ◆ **Pavlopetri – Greece**

The city of Pavlopetri was located off the coast of Laconia in southern Greece. It dates back to around 3,000 BCE and flourished during the Bronze Age. Submerged due to seismic activity and rising sea levels, remains of streets, houses, tombs, and courtyards are still visible, making it the oldest known underwater city in the world.

♦ **Dwarka – India**

Dwarka, located off the coast of Gujarat, India, in the Arabian Sea, is believed to date back as far as 1500 BCE or earlier. Possibly submerged by rising sea levels or tectonic activity, its ancient walls, pillars, and structures can still be observed.

♦ **Olous – Greece**

Olous flourished between the Minoan period (3500 BCE) and Roman times, near the island of Crete. Now submerged off the coast of Elounda, most likely caused by earthquakes and rising sea levels, the remnants of walls, buildings, and ancient artifacts have been discovered underwater there.

♦ **Atlit-Yam – Israel**

Atlit-Yam off the coast of Haifa, Israel, in the Mediterranean Sea, dates back to around 7,000–5,000 BCE. Submerged due to rising sea levels after the last Ice Age, this prehistoric settlement had stone houses, graves, and megalithic stone circles, providing crucial insights into early human life.

These ancient sunken cities provide invaluable insights into lost civilizations, their daily lives, and the catastrophic events that led to their demise. While most of the above do not fit in with Plato's timeline of Atlantis, what they do tell us is that many cities, civilisations, and settlements have succumbed to rising sea levels, floods, and seismic activity over the millenia. With that in mind, it wouldn't be too far-fetched to consider that a city such as Atlantis could also have come to the same fate.

Researchers continue to use advanced underwater archaeology to explore these submerged ruins and uncover more about our ancient past. Perhaps, one day, they will discover the mysterious city of Atlantis itself!

What we know and still don't know about Atlantis

The story of Atlantis has fascinated people for centuries, and while much has been debated and speculated, I hope you have enjoyed reading about some of the main theories surrounding the myth.

What we do know for a fact is that the earliest and most detailed mention of Atlantis comes from the ancient Greek philosopher Plato. Although his account may be historical in nature, many scholars prefer to interpret Atlantis as an allegory, a cautionary tale invented by the philosopher to illustrate the dangers of moral decline. In that sense, Atlantis may not represent a real place at all but a metaphor for societal failure and the corrupting influence of power.

The legend of Atlantis has influenced many cultures and thinkers throughout history and has become a symbol of lost civilizations. Many link it to other myths like the Great Flood and stories of lost golden ages across different cultures (e.g., the Minoans, Sumerians, or the legendary city of Tartessos).

Numerous hypotheses link Atlantis to real-world locations and various speculative theories place Atlantis in these regions based on geological evidence, underwater ruins, and ancient maps. Each theory presents different arguments, though none have been definitively proven.

Despite extensive research and explorations, we still have no clear archaeological evidence of Atlantis as described by Plato. Is it still waiting to be discovered in the silent depths of some distant sea?

Since there is no consensus about where Atlantis might have been located, it could be almost anywhere. Some researchers believe that if Atlantis did exist, it might be a misremembered or exaggerated account of real events, like the destruction of Thera or other ancient cultures lost to natural disasters.

If, indeed, Atlantis was real, what kind of society was it? Plato describes it as advanced, wealthy, and powerful, with technologies and social structures that far surpassed others of its time. Some modern theories suggest it was a highly advanced civilization, perhaps even possessing sophisticated technology, but there's no hard evidence as yet to support this claim.

And theories about ancient aliens or advanced lost technologies are in abundance, yet the evidence and scientific validation for such ideas remains to be produced. Perhaps, in time, new archaeological technologies, such as ground-penetrating radar, satellite imagery, and deep-sea exploration, will successfully unveil previously unknown sites. Discoveries like the sunken ruins of ancient cities in places like Doñana National Park or underwater structures in the Bimini Islands may provide new insights or clues about the origins of the Atlantis story.

As for Atlantis' link to Egypt, there doesn't appear to be any concrete archaeological evidence in Egypt that directly refers to Atlantis or confirms that Egyptian priests had historical knowledge of a real sunken civilization resembling Atlantis. Of course, that doesn't mean there is no link: it could simply mean that we haven't discovered the connection yet. While many researchers suggest that the advanced technology or knowledge Plato attributed to Atlantis might resemble the impressive engineering feats of Egypt, such as the pyramids, more physical evidence is needed to convince the mainstream scientific community that these achievements are due to an Atlantean civilization.

As technology improves, researchers may find previously overlooked clues in Egypt's vast written records or inscriptions that provide more insight into early Egyptian perspectives on distant lands and lost civilizations. Until then, Egypt's role in the Atlantis story will have to remain a secret buried in time.

What does the future hold for Atlantis?

If Atlantis was inspired by a real place, the future could hold key discoveries that offer clearer answers about the nature of the civilization that might have inspired Plato's myth. Advances in genetic research and paleoenvironmental studies could shed light on ancient civilizations and their connections. Understanding how ancient populations migrated, adapted, and were impacted by natural disasters might provide even more context for the Atlantis myth.

Likewise, as we continue to study ancient volcanic eruptions, tectonic shifts, and sea-level changes, the findings might reveal more about the catastrophic events that could have inspired stories of lost civilizations like Atlantis.

Atlantis will likely remain a popular topic in both scholarly and speculative research, with alternative history researchers, New Age spiritualists, and esoteric thinkers continuing to explore Atlantis in their efforts to develop theories about human origins, ancient wisdom, or extraterrestrial contact.

The future may lead us to uncover new evidence that either confirms or debunks these theories, with the potential to reshape our understanding of ancient human history. Now, that would be something!

Whatever the future holds, Atlantis continues to fuel our imagination, serving as a timeless reminder of the lost knowledge and civilizations that could lie beneath the surface of history.

What do I believe?

In my formative years, I was equally drawn to the mysteries of the universe's creation and the nature of human life. Immersing myself in these mysteries required navigating the study of cosmology, quantum physics, Vedic literature, and other esoteric sources. The more I studied, the more I became acutely aware that science did not offer any reliable answers to 'source' questions. The almost analogous study of mysticism and esoteric literature led me to question whether, although they provided profound explanations to source questions (which could never be tested), they consistently failed in their predictions about the observable world. This left me with a certain degree of inner restlessness.

It is often said in spiritual circles, "When the student is ready, the teacher appears." And sometimes, the teacher is not a human being but an actual experience. I experienced exactly that.

Through a very unlikely and coincidental series of introductions and meetings with clients and friends across the UK and US, I ended up having dinner with a Romanian music teacher. She had trained in Vienna and was currently living in London, hoping to join a professional music group. She had visited India previously and was deeply interested in Indian spiritual traditions, which is why a mutual friend had suggested we meet. As we spoke that evening, I felt a clear mutual attraction, as would be understandable between two single people in their early 20s with a shared interest.

We met again the following week and spent the entire day together. Before our very long 'date' ended, she shared with me that she was going to Latvia to perform in a couple of weeks and invited me to join her to watch her live performance. She added that her parents, who were EU diplomats, had a nice apartment in the city center where I could stay with her to spend the weekend exploring the city together. I immediately said yes, agreeing to meet her directly in Riga, flying in from Amsterdam, where I would be for business earlier that week.

In Riga, the concert was lovely, we enjoyed exploring the city, and as we were walking down one of the streets, we noticed a sign on a door that said, "Mystic Seer: Revealing the Future with the Third Eye." Amused, we decided to go in. She went first and received the usual predictions about "love and money" coming into her life soon. She was pleased! Then it was my turn. I asked the seer what made his third eye special.

"It can see through the veil of the past," he responded. I almost jokingly asked if he could then tell me about my past, as that would interest me more than the future — which I preferred to keep unknown.

In the next 15 minutes, the seer spoke as if in a trance. I recount below my approximate dialogue with him:

Seer: "Good to see you, learned one. It's been a while since we spoke."

Me: "How do I know you?"

Seer: "I was a king. You were my mentor and advisor in a lifetime we shared."

Me: "Why do I feel a strong sense of déjà vu?"

Seer: "Because you were my teacher. We spent a lot of time talking about everything. I owe you a lot."

Me: "Interesting. But why am I here again?"

Seer: "Because you didn't finish. You taught me the path of wisdom, but you failed to educate me about the pitfalls of power. I wanted to endow your academy, but everything ended too soon."

Me: "Where was this? In Greece?"

Seer: "No, it was a different root race. It was Atlantis."

Me: "What is the purpose of our meeting now?"

Seer: "Last time, I couldn't repay my debt. This time, I'm here just to remind you of who you were and that soon, you must bridge the lessons of Atlantis."

Me: "How will I know when to begin?"

Seer: "It will be after your first trip to Atlantis. You'll see that the end of this epoch is coming, and you'll be ready to establish your new school."

Me: "But I'm no teacher, and I have no clue where Atlantis is."

Seer: "Don't worry, my master. Both the teacher in you and Atlantis will find you."

Needless to say, this was followed by stunned silence. Eventually, I just got up and left. My friend and I remained quiet for the rest of the evening, avoiding eye contact.

The next morning, we flew back to London. Clearly, the experience had been too much for her. We never went out again, though we occasionally ran into each other at social events before eventually losing touch altogether.

Recently, she found me on Instagram. She now lives in Prague with her husband and two kids.

Her first question to me when we reconnected was, "Did you find Atlantis?"

"In the hieroglyphs of ancient Egypt lie the keys to understanding the sacred mysteries of existence, whispers of wisdom that transcend time and space, awaiting the seeker who dares to unlock their profound secrets."

— Cynthia McKinney

4. The Quest for Lost Knowledge: Hieroglyphs, Scrolls, and Sacred Texts

For centuries, the mysteries of ancient Egypt have captured the imagination of scholars, explorers, and dreamers alike. Its towering pyramids, enigmatic Sphinx, and intricate hieroglyphs whisper of a civilization that once possessed profound wisdom and knowledge — knowledge that may have been lost or deliberately hidden from modern understanding.

We are all familiar with hieroglyphs, papyrus scrolls, and the iconic symbols of this great civilization, but do we truly understand their meanings and origins? In this chapter, we embark on a journey through Egypt's sacred texts and mysterious writings, uncovering the secrets locked within these ancient chronicles. What truths did the Egyptians know about the cosmos, the soul, and the fabric of existence itself?

As we delve into these mysteries, we are left to wonder: could the lost wisdom of Egypt hold the key to understanding not only their past, but our future as well? What truths will be revealed to us once we are able to decode these ancient symbols and texts, and how will that shape our perception of our journey here on earth? It's an exciting quest that I am eager to share with you in what follows below.

"Hieroglyphics remain one of history's greatest enigmas, a complex blend of pictorial art and written language that encapsulates the essence of ancient Egyptian civilization. Each symbol is a key to understanding their beliefs, culture, and history."

— Toby Wilkinson

4.1 Decoding the Language of the Gods

To be fair, even the most learned of contemporary scholars are still unable to decipher all of the Egyptian hieroglyphics they have studied. Understanding them is a complex endeavor, and while significant progress has been made since the deciphering of the script in the 19th century, there are still limitations and gaps in our knowledge.

The hieroglyphs that have been successfully deciphered reveal a wealth of information about ancient Egyptian religion, culture, administration, and daily life. From our understanding, hieroglyphs were used primarily for religious texts, monumental inscriptions, and official documents. The key elements of the script — logograms (symbols representing words), phonograms (representing sounds), and determinatives (clarifying meaning) — are generally understood, allowing scholars to read many texts with relative accuracy. Having said that, their complex nature has allowed for rich and nuanced expression but also made the script challenging to decipher.

Well-known texts, such as the *Book of the Dead*, royal inscriptions, and monumental texts found in temples and tombs, have been extensively studied. This has allowed scholars to compile dictionaries and grammars of hieroglyphic language, facilitating the reading of these important works.

But there are limitations and challenges. Not all hiero-glyphs have been definitively identified or understood. Some symbols may have had specific meanings that remain unclear, especially less common or unique signs. Additionally, certain words or phrases may be context-de-pendent, complicating their interpretation.

Hieroglyphs often carry multiple meanings, and the context in which they are used can significantly affect their inter-pretation. The subtleties of religious texts, poetry, and administrative documents may be difficult to fully grasp, es-pecially when cultural references are not fully understood.

Many ancient texts have been lost to history, and variations in dialects or regional practices can further complicate un-derstanding. Hieratic and Demotic scripts, which developed from hieroglyphs for more practical writing, add layers of complexity that researchers continue to study.

The Book of the Dead

The Book of the Dead is one of the most famous collections of ancient Egyptian funerary texts, consisting of spells and prayers intended to assist the deceased in the afterlife. It provides insights into Egyptian beliefs about death, the afterlife, and the gods. Variants of this text exist, reflecting different practices and beliefs across regions and periods. While we have made significant progress in understanding much of what is written, it is unlikely that we fully com-prehend every aspect of this ancient Egyptian funerary text. There are several reasons for this, including cultural, religious, and linguistic complexities, as well as gaps in our historical knowledge.

We do know that the Book of the Dead, also known to the ancient Egyptians as the *Book of Coming Forth by Day,* is a collection of spells, prayers, and incantations designed to help the deceased navigate the afterlife. Its primary purpose was to guide the soul (or *ka*) through the challenges of the underworld, ensuring a safe passage to the afterlife and eternal life with the gods. The spells offer protection, provide knowledge for specific tasks (like reciting the correct names of gods and door guardians), and help the soul avoid the dangers of the underworld.

But the text is not a single, unified book. It is a compilation of spells that evolved over centuries. Different versions existed across time, and individual manuscripts were often customized for the deceased, reflecting personal religious beliefs or local traditions. Some key components, such as the "Weighing of the Heart" against the feather of Ma'at (representing truth and justice), are fairly well-understood due to their prominence in Egyptian religious thought.

Rich in visual symbols, hieroglyphs, and depictions of gods and mythological creatures, the Book of the Dead has been examined closely by scholars, who have managed to interpret many of these symbols. Connections can be made with a broader Egyptian cosmology, such as Osiris's role as the god of the afterlife, Ra's journey across the sky, and the symbolism of rebirth and regeneration.

So, while we understand the general religious framework, there are cultural and theological nuances that may escape us. Ancient Egyptians had a deep, complex belief system regarding death, the soul, and the afterlife, often involving layers of metaphor, symbolism, and spiritual meaning. Some of the specific references within the Book of the Dead may allude to cultural practices, religious rituals, or mythological stories that have been lost to time. Therefore, we cannot fully grasp all the esoteric or mystical meanings embedded in the text as yet.

Since the Book of the Dead also existed in numerous versions, no two manuscripts are exactly the same. Spells were added, adapted, or omitted based on the individual's needs, social status, or beliefs. This variability makes it difficult to claim a complete understanding, as we often study a composite of different versions rather than a definitive "canonical" version.

Certain words and phrases also remain ambiguous. Ancient Egyptian was a highly symbolic language, and some expressions may have double meanings, metaphorical layers, or contextual interpretations that we may not fully appreciate. For instance, certain terms related to the afterlife or spiritual entities may have specific connotations that have been lost.

The text draws from a vast mythological landscape that we don't fully understand either. Egyptian myths were highly regional and could vary in interpretation depending on the period and location. Some of the gods and mythological creatures referenced in the Book of the Dead might not correspond exactly to deities we are familiar with today, leaving room for speculation about their roles or significance.

Did the Book of the Dead contain deeper, more mystical knowledge that has yet to be uncovered or fully understood? Some scholars believe that certain spells might represent not just religious guidance but also profound spiritual knowledge, cosmology, or even astronomical understanding. Perhaps the text carries layers of meaning meant only for the initiated, which evade most modern readers.

How were these texts recited during burial rites? Were there elements of performance or interaction with other ritual objects that we don't fully understand today? Some portions of the text could be parts of a broader ritual, the significance of which has been lost over millennia.

What is clear is that the *Book of the Dead* was not static; it evolved over the New Kingdom and Late Period, with new spells added and old ones adapted. This raises questions about how interpretations of the afterlife, the gods, and the soul may have changed over time. Do variations between manuscripts reflect regional differences in belief or hidden theological debates? That's a question we should reflect upon.

The Pyramid Texts

Also among the oldest religious texts in the world, the *Pyramid Texts* are inscriptions found in the walls of pyramids from the Old Kingdom (circa 2400–2300 BCE). As far as we can tell, they consist of spells intended to protect the pharaoh in the afterlife and facilitate his resurrection. They are inscribed on the walls of the pyramids of several pharaohs in Saqqara, most notably those of Unas, Teti, Pepi I, Merenre, and Pepi II. Unlike the Book of the Dead, which was intended for a broader group of the deceased (including nobility and commoners), the Pyramid Texts were exclusively reserved for royalty.

With these texts, the pharaoh was ensured of a safe journey into the afterlife and to help him achieve a divine status, becoming one with the gods, particularly the sun god Ra and the god of the dead, Osiris. The spells and incantations offer protection from dangers in the underworld, assist the king in navigating the afterlife, and grant him powers necessary for immortality.

The texts consist of over 700 individual spells or utterances, often grouped by theme. Some are directed at ensuring the king's resurrection, while others provide protection from enemies, both human and supernatural. There are spells that invoke the gods to support the king's journey and guarantee his place among them in the afterlife.

Many spells invoke the myths of Osiris and Isis, Ra, and other deities central to Egyptian religion. The pharaoh is often equated with Osiris, who is resurrected after death, symbolizing the cycle of life, death, and rebirth. This alignment with Osiris helped secure the pharaoh's transition from mortal king to divine being.

The texts reflect the ancient Egyptian view of the cosmos and the afterlife. The journey of the soul, or *ba*, was fraught with dangers, but through the power of the spoken word (spells) and proper rites, the soul could ascend to the heavens, join the sun god Ra in his celestial journey, and live eternally. Some spells reflect complex ideas about how the king's body would be reanimated or how his spirit would travel to the sky to join the gods in the afterlife. In the worldview of ancient Egypt, the well-being of the state and its people depended on the successful transition of the king to the afterlife.

Modern scholars have largely deciphered the Pyramid Texts, building on the breakthroughs achieved after the Rosetta Stone. Most of the spells and their meanings are now understood, providing deep insights into Old Kingdom religious practices, views on death, and cosmology. Having said that, even though the core messages of the Pyramid Texts are clear, there are still nuances that may escape modern understanding. This is because ancient Egyptian religious texts often carry layers of metaphor and symbolism. While we understand many of the references to gods, afterlife concepts, and resurrection, there are deeper theological meanings and mythological allusions that still remain obscure.

It has been suggested by some that the Pyramid Texts may contain hidden or esoteric knowledge, reserved only for the initiated priests or those well-versed in Egyptian theology. There may be deeper spiritual or mystical meanings embedded within the spells, beyond the literal interpretation of the king's journey to the afterlife.

The Pyramid Texts reflect early Egyptian religious thought, which evolved significantly in later periods. How these early beliefs transitioned into the religious practices of the Middle and New Kingdoms, and whether some knowledge was lost or transformed, remains an area of research. Did some of the cosmological and religious ideas in the texts change over time, or were they preserved in a more mystical, hidden tradition?

The placement of the texts within the pyramids themselves — on the walls of burial chambers and corridors — raises questions about their ritual function. Were they meant to be recited by priests during the burial ceremonies, or did the mere presence of the written words carry magical power? While some of these practices are known, the full ritual context of the texts' usage may remain unclear.

Considering that the Pyramid Texts are a crucial key to understanding the religious and cosmological worldview of the Old Kingdom of Egypt, the symbolic depth of these ancient spells leaves some questions unanswered. As we continue to study the texts in their archaeological context, there remains the potential to uncover new insights into ancient Egyptian beliefs and practices, as well as the role of kingship.

The Hieratic Book of the Dead, the Demotic Chronicle, and the Lepsius Papyrus

The *Hieratic* version of the Book of the Dead was a practical adaptation of the original text, written in cursive script on papyrus, often customized with different spells for individual burials. Similarly, the *Demotic Chronicle*, an important historical document from the late periods of Egyptian history, particularly the Ptolemaic era, offers insights into the socio-political context of the time but can be difficult to interpret due to variations in dialect. The *Lepsius Papyrus*, named after Egyptologist Karl Richard Lepsius, contains legal and administrative texts in Demotic, providing valuable glimpses into the daily life, economy, and legal systems of ancient Egypt. Together, these texts reflect the complexities of maintaining cosmic and social order in Egyptian culture.

Unfortunately, the existence of these different scripts — hieroglyphic, Hieratic, and Demotic, each with its own context and usage, adds layers of complexity to our understanding of the texts. Different regions or periods may reflect local dialects or customs, complicating the translation and interpretation process. For instance, the language used in a temple inscription may differ significantly from that found in a legal contract.

Many ancient texts have also been lost to time due to decay, destruction, or neglect. This loss creates gaps in our complete understanding of the full spectrum of Egyptian thought, culture, and practices. In addition, the nuances of religious, legal, and daily life texts require careful interpretation, as they may employ symbolism or cultural references that are not immediately clear to us modern readers.

Researchers face great challenges in decoding the layers of meaning embedded within hieroglyphs, Hieratic, and Demotic scripts. Each text is a fragment of a larger narrative, offering insights into a civilization that was both intricate and profoundly influential in human history. We do not fully understand the meaning of everything they left us on dusty papyrus and in sacred temples or pyramids. It is because of his lack of understanding that questions still arise — questions about the wisdom that the ancient Egyptians held knowledge of, questions that we need to keep asking!

Many hieroglyphic texts contain mystical or esoteric knowledge that historians, archaeologists, and researchers find difficult to understand, even if some of the symbols themselves are decipherable. Since religious concepts and symbolism can be deeply rooted in cultural beliefs, they are often challenging to interpret for the modern mind, which is often trapped in the western notions of scientific inquiry alone.

We also have to acknowledge that our understanding of such ancient texts and symbols has been skewed by early historians, who based their interpretations of the writings on their own subjective, cultural perspectives. As is usually the case, we inherit 'words of wisdom' by those writing the history books, who are very often influenced by their own prejudices, interests, and lack of expertise. While this doesn't mean they are necessarily wrong, it does open up the possibility that something has been missed along the way, lost in translation.

What if the ancient inscriptions on the walls of ancient Egypt held deeper, esoteric meanings that were ignored or misinterpreted? What if the lost scrolls and papyrus that have come to light over the centuries contain within them mystical secrets that we still cannot fully decode? It's an exciting thought!

What Did the Greeks Think?

The ancient Greeks had a complex relationship with Egyptian hieroglyphics, which evolved through different periods of interaction, exploration, and cultural exchange. Naturally, the Greeks were fascinated by the grandeur of ancient Egyptian civilization, including its monumental architecture, religious practices, and wisdom. This admiration is evident in the writings of early Greek historians and philosophers, such as Herodotus, who traveled to Egypt in the 5th century BCE. He described the Egyptians as a highly advanced people, attributing various philosophical and scientific ideas to their culture.

The Greeks often viewed hieroglyphics as mystical symbols rather than a writing system. The term "hieroglyph," is a Greek word, meaning "sacred carving," which reflects this perception. They believed the symbols had divine significance and thought that hieroglyphs contained hidden meanings and were associated with religious practices. What is interesting is that most modern-day archaeologists prefer to refute this idea, maintaining that hieroglyphs are nothing more than a complex writing system, even though the word 'hieroglyph' itself contains the essence of something more esoteric.

As Greek culture expanded, especially during the Hellenistic period following Alexander the Great's conquest of Egypt in

332 BCE, there was a significant exchange of knowledge. The Greeks encountered the wealth of Egyptian learning, which included mathematics, astronomy, medicine, and philosophy. Greek scholars such as Manetho, an Egyptian priest and historian, began to document Egyptian history in Greek, integrating their own interpretations. Although they often lacked a full understanding of hieroglyphs, they recorded Egyptian traditions and religious beliefs, influencing later Greek thought.

The Rosetta Stone

During the Ptolemaic dynasty (305–30 BCE), when Greek rulers governed Egypt, there was a concerted effort to translate and interpret Egyptian texts. The Rosetta Stone, created in 196 BCE, reflects this blending of cultures, featuring Greek text alongside hieroglyphics and Demotic script.

This granodiorite stele was inscribed with a decree in three different scripts:

- **Hieroglyphic**: The upper section contains the text in hieroglyphics, the script used for religious and monumental purposes.

- **Demotic**: The middle section is inscribed in Demotic, which was the common script of Egypt at the time, used for everyday writing and administration.

- **Greek**: The bottom section features the text in Greek, which was well-known to scholars at the time of the stone's discovery. This provided a crucial key for understanding the other two scripts.

The stone was discovered in 1799 by French soldiers in the town of Rosetta (modern-day Rashid) during Napoleon's

campaign in Egypt. Its significance lies in its role in the decipherment of Egyptian hieroglyphs, which began in 1822 by the French scholar Jean-François Champollion. He was able to successfully use the Greek text as a foundation to decode the hieroglyphs on recognising that hieroglyphs included both phonetic and symbolic elements. This allowed him to translate many of the symbols and understand their meanings.

The decipherment of the Rosetta Stone opened the door to the study of ancient Egyptian language, culture, and history, significantly advancing the field of Egyptology. It also underscored the importance of translation and interpretation in understanding ancient civilizations. The question we can ask here is: was the Greek translation of the hieroglyphics reliable? If so, then we can rely on the knowledge it reveals to us. But if it was tainted unintentionally by cultural, historical, and even political biases, then how much of the translation can we credibly accept?

Deciphering the Hieroglyphs

Some Greek scholars attempted to decipher hieroglyphs and related them to their own alphabet and sounds. However, without the knowledge of the phonetic nature of hieroglyphs, these early efforts were largely speculative and often inaccurate.

Indeed, the Greek fascination with Egyptian hieroglyphs and their mystical interpretations influenced later philosophical and esoteric traditions. For example, Hermes Trismegistus, became emblematic of the blending of Greek and Egyptian wisdom, often associated with alchemy and mysticism.

The Greeks viewed hieroglyphics through a lens of both reverence and confusion, admiring the sophistication of

Egyptian civilization while grappling with their understanding of the hieroglyphic writing system. Their interactions with Egyptian culture laid the groundwork for later interpretations and scholarship, influencing not only their own intellectual traditions but also the Renaissance revival of interest in ancient Egypt.

Ultimately, the Greek perspective on hieroglyphs reflects a complex interplay of admiration, mystery, and the pursuit of knowledge that transcended cultural boundaries. At the same time, we cannot know what gaps in their knowledge they attempted to fill with subjective interpretations and cultural bias. For instance, when they encountered concepts or deities that had no direct equivalents in Greek thought, they sometimes projected their own beliefs onto these Egyptian ideas, leading to a distorted understanding.

The Evolution of Understanding

Throughout history, our understanding of hieroglyphs has evolved significantly. Before Champollion, scholars often viewed hieroglyphs as mere pictograms or symbols representing ideas rather than a fully developed writing system. This led to numerous misinterpretations and fanciful theories about their meanings.

With the deciphering of hieroglyphs, scholars began to recognize the linguistic sophistication of ancient Egyptian. Hieroglyphs are now understood as part of a broader continuum of the Egyptian language, which evolved through various stages: Old Egyptian, Middle Egyptian (the classical language of hieroglyphs), Late Egyptian, and Coptic.

Through understanding them, we can garner invaluable insights into ancient Egyptian religion, society, and governance. Today, researchers continue to study hieroglyphs, employing new technologies such as digital imaging and artificial intelligence to analyze ancient texts and artifacts. This ongoing research promises to uncover even more about the rich tapestry of ancient Egyptian civilization. It may even reveal new secrets hidden within the symbols – secrets that will redefine our understanding of the timeless wisdom of the ancient Egyptians.

"The discovery of lost scrolls can revolutionize our understanding of ancient Egyptian beliefs and practices. These texts often hold esoteric knowledge that was meant to be safeguarded, hinting at deeper spiritual truths."

— Salima Ikram

4.2 Lost Scrolls and Forbidden Wisdom

Lost scrolls are a significant part of the mystery surrounding ancient Egypt's intellectual and spiritual history. These scrolls, many of which were written on papyrus, contained a wealth of knowledge — ranging from religious and magical texts to scientific treatises on medicine, mathematics, and astronomy — that are now either lost or exist only in fragments.

Some key examples of lost Egyptian scrolls include:

1. **The Library of Alexandria:** One of the most famous repositories of knowledge in the ancient world, the Library of Alexandria, is believed to have housed countless Egyptian scrolls. Many scholars and scribes from Egypt and other cultures contributed to this collection. However, due to its destruction by fire — likely on multiple occasions — many of these texts were lost forever. The knowledge within may have contained invaluable insights into ancient Egyptian history, science, religion, and esoteric thought.

2. **Lost Religious Texts:** While some religious texts like the *Pyramid Texts, Coffin Texts,* and *Book of the Dead* have been preserved, others were likely lost due to the fragility of papyrus and the selective copying over generations. Priests kept sacred

scrolls in temple libraries, and many texts, including potentially esoteric or secret knowledge known only to the religious elite, may have been destroyed or decayed over time.

3. **Scrolls of Thoth:** In Egyptian mythology, Thoth was the god of wisdom, writing, and knowledge, and legends speak of sacred scrolls attributed to him. These *"Books of Thoth"* were said to contain powerful magical knowledge, such as controlling the elements and communicating with gods. It is difficult to say if these scrolls ever physically existed or if they may have been symbolic, but they have fueled speculation that some profound wisdom from ancient Egypt has been lost to history.

4. **Medical and Scientific Scrolls:** Though documents like the *Ebers Papyrus* and *Edwin Smith Papyrus* provide valuable insight into ancient Egyptian medicine, they likely represent only a fraction of the medical knowledge that once existed. Medical, mathematical, and astronomical scrolls were crucial for everyday life in Egypt, and many of these practical texts have vanished over time. What remains, like the *Ebers Papyrus*, suggests a rich tradition of empirical knowledge that is now incomplete.

But how did such scrolls come to be lost? Were they perhaps hidden, stolen, or destroyed intentionally? Many theories abound about the reasons why so many scrolls do not exist today, some of which seem logical, and others, rather suspicious.

- ◆ **Material Decay:** Papyrus, the material on which most scrolls were written, is highly vulnerable to decay, especially when exposed to moisture or poor preservation conditions. It makes perfect sense, then, that many scrolls that survived millennia in Egypt's dry desert climate eventually deteriorated when they were moved or stored improperly.

- ◆ **Destruction and Looting:** Egypt's long history of invasion, colonization, and war led to the destruction and looting of many libraries, temples, and tombs. Scrolls stored in these places were either deliberately destroyed or lost due to neglect.

- ◆ **Selective Preservation:** Not all texts were deemed worthy of copying or preserving. Priests and scribes may have also intentionally let certain knowledge die out, either because it was seen as obsolete or because it contained sacred information meant only for a select few.

Today, archaeologists and Egyptologists continue to search for lost scrolls in unexplored or newly discovered tombs, temple ruins, and ancient libraries. Every new discovery has the potential to reshape our understanding of ancient Egypt and recover some of the knowledge that was thought to be lost forever.

The lost scrolls of ancient Egypt could indeed hold significant insights into the esoteric beliefs and mystical traditions of the civilization. Many scholars and enthusiasts of Egypt's ancient culture speculate that certain texts may have been hidden or destroyed deliberately to protect sacred wisdom, preserving it for only the initiated or to prevent it from falling into the wrong hands.

Esoteric Beliefs in Ancient Egypt

Ancient Egyptian spirituality was deeply esoteric, intertwining religious ritual, magic, and a profound understanding of the cosmos. The priests, particularly those connected to the temples of powerful gods like Ra, Osiris, and Thoth, were believed to possess secret knowledge about the workings of the universe, the afterlife, and the divine order, known as *ma'at*. This sacred wisdom would have been preserved in texts and inscriptions, many of which we do not have in our possession today.

1. **The Mysteries of the Afterlife:** Texts like the *Pyramid Texts, Coffin Texts,* and *Book of the Dead* are filled with complex symbols and instructions designed to guide the deceased through the afterlife. However, it's believed that there were even more mystical teachings regarding death and resurrection, hidden away in secret scrolls or only passed orally between priests.

2. **Cosmic and Astronomical Knowledge:** Ancient Egyptians possessed an advanced understanding of the stars, planets, and celestial movements, which were deeply tied to their religious beliefs. The temples, like those at Dendera and Abu Simbel, were aligned with celestial events. It's possible that lost scrolls could contain esoteric astronomical or astrological knowledge, which the priests used to maintain cosmic balance.

3. **Magical and Healing Practices:** Ancient Egyptian texts often combined practical medicine with spiritual and magical elements. Scrolls like the *Ebers Papyrus* reveal some of this knowledge, but it is

believed that there were secret magical texts, too, which could have contained spells, incantations, and rituals for powerful magic, healing, and protection.

Here's a good question we should consider: Why might scrolls have been hidden or destroyed?

Many believe that some were deliberately hidden to protect sacred knowledge from the general public or invaders. Esoteric teachings were often restricted to high-ranking priests or members of the royal family, who were believed to have the necessary spiritual purity and authority to handle such wisdom. If these texts contained information about magical practices, astral knowledge, or divine secrets, the Egyptians may have feared their misuse.

During times of invasion, particularly by the Persians, Greeks, and later the Romans, the Egyptian priesthood may have chosen to destroy or hide certain scrolls to prevent the conquerors from gaining access to sacred wisdom. This could be a reason for the loss of many important religious texts and magical manuscripts.

Much like other ancient mystery schools, the Egyptian religious tradition likely included a form of initiation. Certain texts may have been reserved for initiates who had undergone specific rituals to receive higher spiritual teachings. These writings, containing the most sacred mysteries of the gods, the soul, and the afterlife, may have been hidden to prevent uninitiated individuals from understanding or corrupting them.

Some examples of texts and the esoteric wisdom that may have been lost include:

1. **Books of Thoth:** These legendary scrolls, attributed to the god Thoth, were said to contain the knowledge of the gods, including spells to control nature, understand the universe, and commune with divine entities. According to Egyptian mythology, these books were hidden because they contained knowledge too powerful for ordinary humans. They may have contained sacred secrets of alchemy, astronomy, and spiritual ascension. The works of 19th- and early 20th-century scholars like Gerald Massey and later writers like Graham Hancock suggest that this hidden knowledge, if rediscovered, could transform modern understanding of ancient technologies and spiritual practices.

2. **Esoteric Mysticism and the Hermetica:** The *"Hermetica"*, a collection of philosophical and spiritual writings draws from ancient Egyptian wisdom. Advocates like Manly P. Hall have pointed to the Hermetica as evidence of Egypt's profound influence on mystical and esoteric traditions, including concepts of divine knowledge, the unity of the cosmos, and human potential for enlightenment. Many believe that the Hermetica is a surviving fragment of much deeper Egyptian wisdom that has been lost over the millennia .

3. **Advanced Scientific Knowledge:** Some proponents of lost Egyptian wisdom argue that ancient scrolls may have contained advanced scientific knowledge, particularly in astronomy, mathematics, and architecture. The precision with which the Egyptians built the pyramids and aligned them with celestial bodies suggests, to these

advocates, a sophisticated understanding of the stars and cosmic cycles. Graham Hancock, in his *"Fingerprints of the Gods"*, suggests that the lost scrolls could reveal a forgotten understanding of Earth's precession and advanced engineering skills .

4. **Healing and Alchemical Secrets:** Medical texts like the *Ebers Papyrus* give glimpses into the advanced medical practices of the Egyptians, but advocates believe much more remains undiscovered. They argue that the Egyptians may have had knowledge of alchemical processes, healing remedies, and even life-extending techniques that were passed down through secret teachings. Advocates often point to the combination of practical and magical remedies in the surviving medical texts, indicating that lost scrolls may contain further understanding of the metaphysical aspects of healing.

5. **The Destruction of the Library of Alexandria:** The destruction of the Library of Alexandria is often cited by advocates as a monumental loss of Egyptian wisdom. The library is believed to have housed countless scrolls on a variety of subjects, including esoteric, scientific, and philosophical knowledge from Egypt and beyond. Authors like Hancock and John Anthony West suggest that if these scrolls had survived, they might contain knowledge about the origins of civilization, advanced technologies, and humanity's spiritual connection to the cosmos

6. **Scrolls in Temple Libraries:** Temple libraries, particularly those at Karnak and Heliopolis, were

known to house sacred texts that may never have been copied or widely circulated. Some scholars believe that during the decline of Egyptian power or during foreign occupation, these libraries were looted or destroyed to protect their knowledge from being misused. The lost wisdom likely encompassed profound esoteric, scientific, and philosophical knowledge that was central to their civilization. These libraries may have housed sacred texts on cosmic and astronomical phenomena, detailed rituals for maintaining cosmic balance (*ma'at*), magical and alchemical practices, and advanced medical knowledge intertwined with spiritual healing. The loss of such scrolls represents a significant gap in our understanding of ancient Egyptian spirituality and wisdom, and their rediscovery could potentially revolutionize our insight into this enigmatic civilization.

7. **Mystical Knowledge of the Pharaohs:** The mystical knowledge of the pharaohs, viewed as divine intermediaries between the gods and the people, likely encompassed esoteric teachings about the afterlife, the nature of the gods, and the cosmic order. This sacred wisdom may have included detailed instructions on navigating the afterlife, rituals for spiritual ascension, and profound insights into the interconnectedness of the universe. Many scholars suggest that scrolls containing such esoteric knowledge were intentionally buried with pharaohs or hidden in their royal tombs to ensure the preservation of these secrets for the afterlife journey. As many of these tombs remain

undiscovered or unexplored, the potential for un-covering lost texts that could reveal the depth of the pharaohs' mystical teachings continues to cap-tivate researchers and enthusiasts alike, promising to shed light on the spiritual practices and beliefs that shaped ancient Egyptian civilization.

The possibility that ancient Egyptians deliberately hid certain scrolls means that there could still be caches of esoteric wisdom waiting to be found. If uncovered, these scrolls could revolutionize our understanding of Egyptian spirituality and the profound mysteries they sought to protect.

Lost Esoteric Secrets

From an esoteric standpoint, hidden knowledge in ancient Egypt holds profound significance as it embodies truths and insights that transcend conventional understanding, offering pathways to spiritual enlightenment, transfor-mation, and deeper comprehension of the cosmos. This knowledge was often deemed sacred and reserved for the initiated — primarily the priestly class — believed to unlock the mysteries of existence, the nature of the divine, and the soul's journey after death.

Let's take a look at some of the key aspects of hidden knowledge in ancient Egyptian esotericism:

1. **The Path to Enlightenment:** Hidden knowledge was integral to personal and spiritual development in ancient Egypt. *The Book of the Dead* contains spells and teachings intended to guide the de-ceased through the afterlife, emphasizing the importance of inner wisdom for achieving spiritual liberation. Understanding these texts was crucial

for navigating the afterlife and attaining eternal life.

2. **Initiation and Mystical Traditions:** Initiation was a significant aspect of accessing hidden knowledge in ancient Egyptian mystery schools. Priests underwent rigorous training to prepare for esoteric teachings related to the gods and the cosmos. *The Emerald Tablet*, often linked to Thoth, suggests that true understanding of the universe and self comes through profound spiritual practice and initiation.

3. **Guardianship of Sacred Teachings:** Many ancient Egyptian texts were protected and only shared with a select few. This guardianship ensured that powerful knowledge remained within the priestly elite, safeguarding it from misuse. For example, texts related to healing and magic, such as those found in the *Ebers Papyrus*, were considered sacred and closely guarded by the priests.

4. **Symbolism and Interpretation:** Ancient Egyptian esotericism frequently employed symbols, hieroglyphs, and metaphors to convey hidden truths. The intricate symbolism in tomb paintings and inscriptions reflects the belief that deeper meanings required contemplation and insight to decode. The symbolism of the scarab beetle, for instance, represented rebirth and transformation, illustrating how hidden knowledge could lead to spiritual renewal.

5. **Connection to the Divine:** Accessing hidden knowledge was seen as a way to establish a

connection to the divine. The concept of *Ma'at*, representing cosmic order and truth, was central to Egyptian belief. Priests and pharaohs were expected to embody and uphold *ma'at*, reflecting the notion that hidden wisdom facilitated a deeper relationship with the gods and the universe.

6. **Alchemy and Transformation:** Alche-mical traditions in ancient Egypt symbolized hidden knowledge as a means of personal transformation. The process of spiritual refinement was metaphorically linked to the transformation of materials, such as the quest for gold. Egyptian alchemy often emphasized the pursuit of inner gold — spiritual enlightenment achieved through understanding the self and the cosmos.

The significance of hidden knowledge in ancient Egypt lies in its potential to illuminate the path to spiritual growth, offer profound insights into the nature of reality, and foster a deeper connection with the divine. As seekers engaged with the intricate layers of meaning embedded in Egyptian traditions, they were invited to explore the mysteries of existence, unlocking transformative wisdom that transcended ordinary understanding.

The Vedic Connection

It may come as a surprise to you to read that notions of hidden knowledge in ancient Egypt resonate deeply with concepts found in ancient Vedic texts, highlighting a shared understanding of spiritual growth and the pursuit of deeper truths. Both traditions emphasize the transformative power of esoteric knowledge and its role in fostering a connection with the divine.

We can draw many parallels between the two, which raises the question of whether or not the ancient Egyptians inherited their esoteric knowledge from the Vedic sages themselves.

1. **Path to Spiritual Enlightenment:** In both traditions, hidden knowledge serves as a vital pathway to spiritual enlightenment. In Vedic texts, particularly the *Upanishads*, seekers are encouraged to explore the nature of the self (*Atman*) and its unity with the universal consciousness (*Brahman*). This quest for self-realization mirrors the Egyptian emphasis on understanding one's journey through the afterlife and the ultimate reunion with the divine.

2. **Initiation and Esoteric Practices:** Similar to the initiation rites of Egyptian priests, Vedic traditions also incorporate initiation rituals (*diksha*) that prepare individuals for deeper spiritual teachings. These rites symbolize entry into a sacred knowledge system that requires guidance from a teacher or guru. This reflects a common belief that profound truths must be approached with respect and preparation.

3. **Symbolism and Interpretation:** Both traditions utilize rich symbolism to convey deeper meanings.

Egyptian hieroglyphs and Vedic metaphors, such as those found in the *Bhagavad Gita,* encapsulate complex philosophical ideas in accessible forms. For example, the symbolism of the lotus in Vedic texts represents purity and spiritual awakening, much like the scarab beetle in Egyptian symbolism signifies rebirth and transformation.

4. **Connection to the Divine:** The pursuit of hidden knowledge in both traditions emphasizes establishing a personal connection with the divine. In Egyptian thought, *Ma'at* represents the cosmic order that one must align with, while in Vedic philosophy, dharma reflects the moral and ethical duties that align an individual with the cosmic law. Both frameworks suggest that understanding these principles is essential for spiritual harmony and divine connection.

5. **Transformation and Alchemy:** The concept of inner transformation is central to both Egyptian and Vedic traditions. In alchemy, both spiritual and physical transformation are seen as interrelated processes. In Vedic texts, the journey of self-realization involves transcending the ego and realizing one's true nature, akin to the Egyptian quest for spiritual refinement through knowledge.

The exploration of hidden knowledge in ancient Egypt and Vedic texts reveals a shared spiritual framework that transcends cultural boundaries. Both traditions invite seekers to delve into the mysteries of existence, unlocking transformative wisdom that illuminates the path to spiritual growth and fosters a deeper connection with the divine.

This may not be a coincidence. The Vedic sages, or rishis, were active primarily during the early Vedic period, which is generally dated from around 1500 BCE to 500 BCE. This period saw the composition of the earliest texts of the Vedas, including the *Rigveda*, which is the oldest of the four Vedas. Some scholars propose that Vedic wisdom may have been transmitted to the ancient Egyptians through various means:

♦ **Trade Routes:** During the time of the Vedic sages, ancient trade routes were established that connected various cultures across Asia, the Middle East, and North Africa. Through these trade interactions, ideas and cultural practices could have been exchanged. The Silk Road and other routes facilitated the movement of not only goods but also philosophical and spiritual concepts.

♦ **Migration and Cultural Exchange:** Some scholars propose that migrations of peoples could have facilitated the sharing of knowledge between the Indus Valley civilization (which preceded the Vedic period) and other ancient cultures, including Egypt. As societies interacted through trade or migration, it would not be unreasonable to assume that they also absorbed one another's spiritual and philosophical ideas.

♦ **Common Indo-European Roots:** Both Vedic and ancient Egyptian civilizations might share common Indo-European roots in certain philosophical concepts. The study of linguistics and ancient languages reveals similarities in terms and ideas that could suggest a shared ancestral understanding of the cosmos and the divine.

- **Esoteric Traditions:** Both cultures developed rich esoteric traditions, where knowledge was passed down through select initiates. It's conceivable that as priests or sages from different cultures interacted — whether through trade, conquest, or migration — they could have shared their esoteric teachings. For instance, Egyptian mystery schools and Vedic initiation rituals share a common emphasis on secrecy and the sacred nature of knowledge.

- **Philosophical Parallels:** Many themes found in Vedic philosophy — such as the concepts of cosmic order (*Rita*), the nature of the self (*Atman*), and the ultimate reality (*Brahman*) — resonate with Egyptian beliefs about the afterlife and the divine order represented by *ma'at*.

These potential pathways for interaction — through trade, migration, shared cultural roots, and esoteric exchanges — could suggest that both cultures influenced one another indirectly. Taking it one step further, I would like to propose that the Vedas were the source of wisdom for the ancient Egyptians — that the Egyptians received their esoteric knowledge from the ancient sages of the Indus valley.

This proposal is a thought-provoking hypothesis that has been explored by various writers and researchers. The idea that the Vedas were a source of wisdom for the ancient Egyptians and that the Egyptians received esoteric knowledge from the sages of the Indus Valley invites exploration into the interconnectedness of ancient civilizations. While this view is not widely accepted in mainstream academia, several writers and researchers have explored similar themes, proposing potential links between Vedic and Egyptian wisdom.

David Frawley has written extensively about the intersections of Vedic and other ancient wisdom traditions. In *"Yoga and Ayurveda: Self-Healing and Self-Realization" (2000)*, he discusses the parallels between Indian and Egyptian spiritual practices, emphasizing the shared insights into the cosmos and divinity. Frawley argues for a common spiritual heritage that informs both cultures.

Swami Vivekananda emphasized the universality of spiritual truths across cultures in his lectures and writings. In *"Raja Yoga" (1896)*, he discusses the philosophical underpinnings of yoga and spirituality, touching on the interconnectedness of various spiritual traditions, including those of ancient Egypt. Although he does not claim direct transmission, his insights support the idea of a shared spiritual landscape .

Rudolf Steiner, the founder of Anthroposophy, explored the concept of a universal wisdom tradition that transcends cultural boundaries. In *"Theosophy: An Introduction to the Spiritual Processes in Human Life and in the Cosmos" (1923)*, he examines the similarities between the esoteric knowledge of Vedic sages and Egyptian priests. Steiner posits that both traditions sought to understand the divine and the cosmos, suggesting potential pathways for knowledge exchange.

In *"Forbidden Archaeology: The Hidden History of the Human Race" (1998)*, Michael Cremo discusses the possibility of advanced civilizations existing earlier than traditionally thought, which may have facilitated knowledge sharing among cultures, including the Indus Valley and Egypt. While he does not explicitly claim that Vedic wisdom influenced Egyptian thought, his ideas about cultural exchanges align with this hypothesis.

R. C. Zaehner's works often explore the interconnectedness of ancient religious traditions. In *"Hinduism" (1962),* he examines the similarities in cosmology and esoteric teachings across cultures, including India and Egypt. His comparative approach suggests that there could have been a flow of ideas and wisdom between these civilizations, supporting the idea of shared insights.

While direct evidence for the claim that the Vedas were the source of wisdom for ancient Egypt is limited, the writings of these scholars and thinkers highlight potential parallels and avenues for cultural exchange between these ancient civilizations. As we do further interdisciplinary research into linguistics, archaeology, and comparative religion, we may be illuminated by what we find out!

Theosophy on Ancient Egypt and the Vedas

The idea that the Vedas were a source of wisdom for ancient Egypt aligns closely with several principles found in Theosophy and the writings of its central figures, Helena Blavatsky and Henry Steel Olcott. Theosophy emphasizes the interconnectedness of all spiritual traditions and the pursuit of universal truths. In relation to Vedic wisdom and ancient Egyptian knowledge, these truths are presented in the works: *"The Secret Doctrine"* (H. P. Blavatsky, 1888) and *"Old Diary Leaves"* (H. S. Olcott, 1880).

1. Universal Wisdom Tradition

Theosophy posits that all major religions share a common core of spiritual truth, which transcends cultural and historical boundaries. This idea supports the notion that the Vedas and Egyptian esoteric teachings may be manifestations of a shared ancient wisdom. Theosophists believe that this sacred 'knowing' can be traced back to a singular source, which resonates with the hypothesis that both civilizations derived insights from a common spiritual heritage.

2. Reincarnation and Karma

Theosophy teaches concepts such as reincarnation and karma, which are also reflected in Vedic thought. The belief in a cyclical view of life and the moral implications of one's actions aligns with Egyptian ideas about the afterlife, judgment, and moral order (*ma'at*). This philosophical overlap suggests that both cultures may have drawn from similar spiritual frameworks, reinforcing the idea of shared wisdom.

3. Esoteric Knowledge

Theosophy places a strong emphasis on esoteric knowledge — wisdom that is hidden or reserved for a select few. The claim that ancient Egyptians received esoteric teachings from Vedic sages aligns with Theosophical views that advocate for the uncovering of hidden spiritual truths. Theosophists believe that understanding these truths can lead to spiritual enlightenment and personal transformation.

4. Spiritual Hierarchies and Masters

Theosophical teachings often reference advanced spiritual beings or "Masters" who guide humanity's spiritual evolution. Some Theosophists argue that these Masters have historically influenced various cultures and their spiritual traditions, including those of the Vedic sages and Egyptian priests. This idea suggests that the transfer of knowledge between cultures could have been facilitated by these spiritual guides.

5. Cultural Exchange and Influence

Theosophy emphasizes the importance of cultural exchange in the development of spiritual knowledge. Proponents of the idea that Vedic wisdom influenced Egyptian thought may argue that ancient trade routes and migrations allowed for the transmission of ideas. This perspective resonates with Theosophical beliefs about the interconnectedness of all cultures and their contributions to the collective spiritual understanding of humanity.

The hypothesis that Vedic thought served as a source for ancient Egyptian knowledge fits well within Theosophical thought, which champions the universality of spiritual truths and the interconnectedness of all human experiences. By exploring these ancient traditions through a Theosophical lens, we may just uncover deeper insights into the shared quest for understanding the divine and the mysteries of existence.

Atlantean Knowledge and Ancient Egypt

The idea that the ancient Egyptians may have been trying to "recollect" and "preserve" Atlantean knowledge is an intriguing hypothesis that has appeared in various esoteric and speculative writings. This concept ties into the broader themes of lost civilizations and ancient wisdom. If you recall from Chapter 3, we talked about Plato's notion of Atlantis, where he describes a highly advanced civilization that existed about 9,000 years before his time. Some theorists suggest that the Egyptians, through their own advancements, could have sought to preserve remnants of this lost knowledge.

They also take the discussion further, arguing that ancient civilizations often maintain a collective memory of previous societies through myths and religious traditions. The Egyptians had a rich mythological system that included references to ancient wisdom and lost knowledge, which could be interpreted as echoes of Atlantean teachings. Is it possible that texts like the Pyramid Texts may contain fragments of this collective memory?

Many esoteric traditions claim that ancient Egyptians were custodians of deep spiritual and philosophical knowledge, akin to the wisdom attributed to Atlantis. Mystery schools in ancient Egypt, which trained initiates in the secrets of life, death, and the cosmos, are sometimes thought to be continuations of Atlantean teachings. This idea is certainly popular in Theosophical literature and among certain New Age thinkers today.

Let's not forget the advanced architecture and engineering feats of the ancient Egyptians, such as the construction of the pyramids, which have led some to speculate about lost technologies that could have been derived from Atlantean knowledge. The precision and scale of these structures raise questions about the sources of their architectural wisdom, potentially linking them to a more ancient civilization.

When we look at the similarities between the mythologies, cosmologies, and spiritual practices of ancient cultures, including Egypt and those attributed to Atlantis, it isn't difficult to imagine that these parallels could indicate a shared source of knowledge or influence. The hypothesis certainly invites further exploration into the connections between ancient cultures, their myths, and their spiritual legacies.

In *"Forbidden Archaeology,"* Michael Cremo discusses the possibility of advanced civilizations existing earlier than traditionally accepted timelines. While he does not specifically focus on Atlantis, his arguments about lost knowledge and ancient wisdom resonate with proponents of the idea that the Egyptians sought to preserve such knowledge. The kind of evidence he and his supporters cite include:

◆ Mythological Parallels: Proponents often cite similarities in myths and religious practices between Egypt and other ancient cultures as evidence of shared wisdom.

◆ Architectural Similarities: They point to architectural feats in Egypt, such as the pyramids, as potentially indicative of lost technologies that might connect back to Atlantis.

◆ Linguistic and Cultural Links: Some researchers look for linguistic and cultural connections between the peoples of the Indus Valley, Egypt, and hypothetical Atlantean traditions.

Although the idea that ancient Egyptians preserved Atlantean knowledge remains speculative and controversial, contemporary work on the subject makes for a fascinating quest for the truth about the lost wisdom of Egypt.

Library of Lost Knowledge

The Al-Hammadi Library, located in the region of Al-Hammadi in Egypt, is often associated with the discovery of significant ancient texts, particularly those related to Gnostic and early Christian thought. While it does not specifically focus on Egyptian wisdom per se, the library's contents and context provide insights into the complex interplay between various spiritual traditions, including those of ancient Egypt.

The library dates back to the 2nd and 3rd centuries CE and is believed to contain texts that reflect the spiritual and philosophical currents of the time, during which ancient Egyptian beliefs were still influential. The library is part of the broader context of early Christian communities that emerged in the aftermath of the decline of traditional pagan practices, including those of ancient Egypt.

Here, we can find Gnostic writings, which often incorporate elements of Egyptian mythology and cosmology. Gnosticism, with its emphasis on hidden knowledge and spiritual awakening, echoes themes found in ancient Egyptian esoteric teachings. For instance, texts such as the *Gospel of Thomas* and the *Apocryphon of John* explore concepts like the divine spark within humanity, which resonates with Egyptian beliefs about the soul and the afterlife.

The Al-Hammadi Library serves as a reminder of the loss of ancient knowledge due to the rise of orthodox religious movements. As Christianity became more dominant, many Gnostic texts and teachings were suppressed or destroyed. The survival of these texts in the Library provides a glimpse into the spiritual landscape of late antiquity, revealing how ancient Egyptian wisdom may have influenced early Christian thought and practices.

The library is also a testament to the interconnectedness of various spiritual traditions in the ancient world. The incorporation of Egyptian themes in Gnostic texts suggests that even as Egyptian civilization declined, its ideas continued to permeate and influence emerging belief systems. This blending of traditions may indicate an attempt to preserve aspects of ancient wisdom within new frameworks.

The texts to be found in the library reflect the continuity and transformation of spiritual knowledge in the ancient world, underscoring the importance of hidden and esoteric teachings that have shaped various religious and philosophical traditions. Such texts echo the broader themes of lost knowledge and the quest for understanding that characterize both ancient Egypt and its successors.

Here are some notable examples of the kinds of esoteric thought found in the texts from the library:

+ The Gospel of Thomas

This collection of sayings attributed to Jesus emphasizes personal spiritual knowledge and direct experience of the divine. One notable saying states, "The kingdom of God is inside you and all around you." This idea resonates with the Egyptian concept of the divine presence within and the importance of inner awakening.

+ The Apocryphon of John

This text presents a Gnostic interpretation of creation and the nature of God, describing a complex cosmology with multiple divine beings. It explores themes of the divine spark within humanity and the importance of gnosis (knowledge) for salvation. The notion of a divine inner light echoes Egyptian beliefs about the soul's journey and the importance of knowing oneself.

+ The Gospel of Truth

Attributed to Valentinus, this text discusses the relationship between knowledge and ignorance, emphasizing that understanding the divine leads to liberation from the material world. The text reflects the Egyptian emphasis on the importance of knowledge in the afterlife, particularly regarding the soul's judgment.

◆ The Book of Thomas the Contender

In this text, Thomas engages in a dialogue with Jesus about the nature of existence, suffering, and the path to spiritual enlightenment. The wisdom expressed here parallels Egyptian teachings about navigating the challenges of life and the quest for spiritual truth.

◆ The Hypostasis of the Archons

This text critiques the material world and the archons (rulers) who govern it, suggesting that true understanding comes from recognizing the limitations of the physical realm. This idea aligns with Egyptian esoteric thought, which often emphasized the illusion of the material world and the pursuit of higher knowledge.

The texts found in the Al-Hammadi Library reflect a wealth of thought that blends Gnostic teachings with elements reminiscent of earlier spiritual traditions, including those of ancient Egypt. These texts invite us to explore themes of inner knowledge, divine presence, and the transformative power of understanding, offering insights that continue to resonate today.

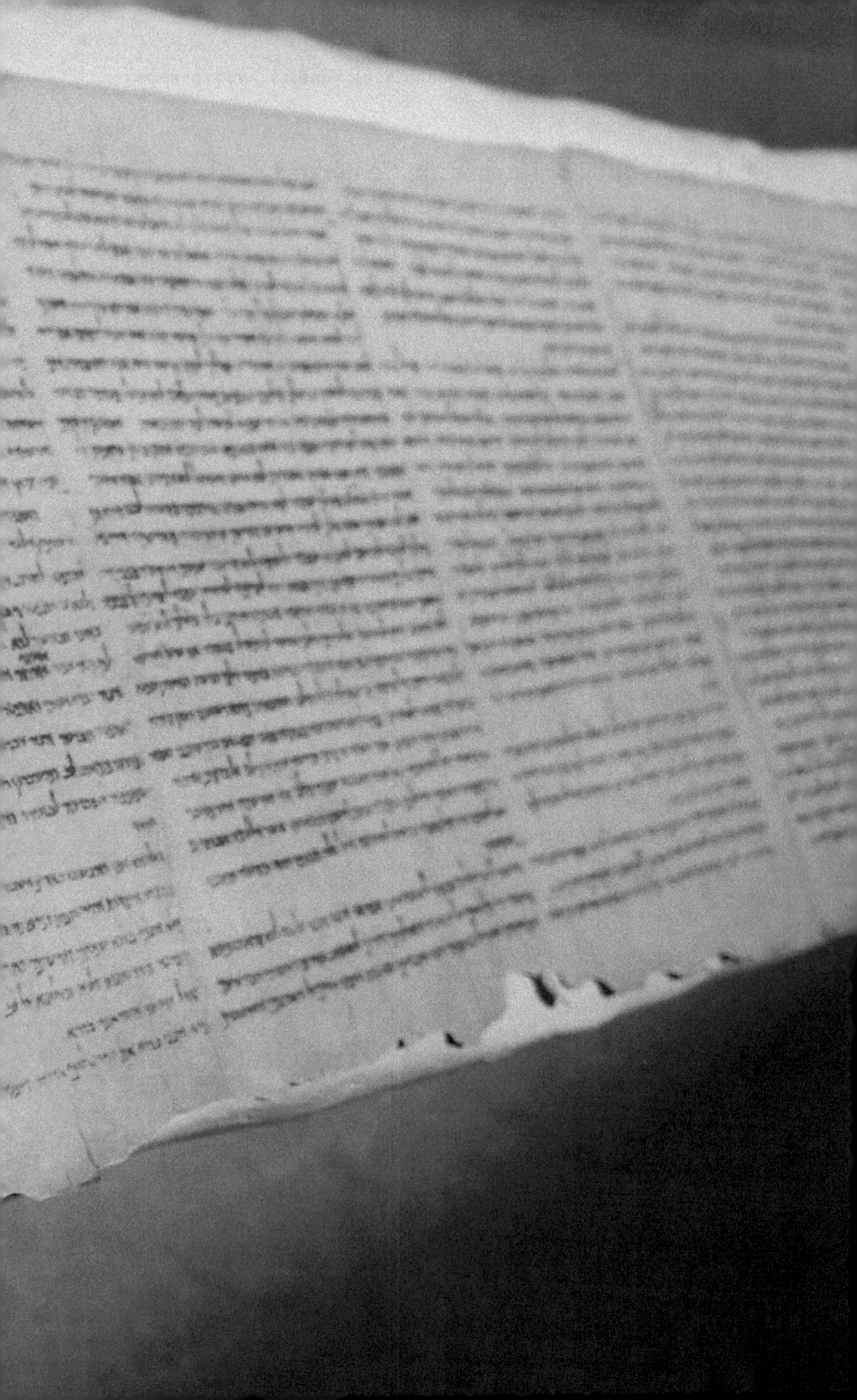

The Dead Sea Scrolls

The Dead Sea Scrolls, discovered between 1947 and 1956 near the Dead Sea, are a collection of Jewish texts that date from the third century BCE to the first century CE. While they primarily reflect the religious beliefs and practices of the Jewish community at Qumran, their connection to the lost wisdom of ancient Egypt can be explored through several intriguing facets:

The period during which the Dead Sea Scrolls were written coincided with significant cultural interactions between Jewish, Hellenistic, and Egyptian societies. This context suggests that ideas may have flowed between these cultures, influencing theological concepts and spiritual practices. For example, some scholars argue that the Qumran community's emphasis on esoteric knowledge and apocalyptic visions may have been influenced by Egyptian mystical traditions.

Certain themes found in the Dead Sea Scrolls also reso-nate with ancient Egyptian thought, particularly regarding concepts of divine judgment, the afterlife, and the pursuit of holiness. For instance, the Scrolls contain discussions about the "Teacher of Righteousness," a messianic figure, paralleling Egyptian ideas about divine kingship and the role of pharaohs as intermediaries between the gods and humanity.

Just as the Egyptians sought to preserve their sacred knowledge through texts like the Pyramid Texts and the Book of the Dead, the Qumran community compiled their writings to safeguard their interpretations of scripture and teachings. This desire to maintain esoteric wisdom high-lights a common thread between the two cultures — the understanding that written knowledge holds power and that it can guide the soul toward enlightenment.

The ideas and texts found in the Dead Sea Scrolls contributed to the development of early Christian thought. As with Egyptian wisdom that permeated later religious practices, the Scrolls reflect a confluence of spiritual ideas that shaped the beliefs of subsequent generations. This interplay illustrates the broader theme of lost knowledge influencing emerging spiritual paradigms.

While the Dead Sea Scrolls are primarily rooted in Jewish tradition, their historical context and thematic parallels suggest a healthy abundance of cultural exchange that could encompass elements of ancient Egyptian wisdom. By examining these connections, we gain a deeper understanding of how lost knowledge may have traversed time and space, shaping the spiritual landscapes of various civilizations and leaving a lasting legacy.

The Hidden Truth in Other Discoveries

There are several other notable discoveries and texts related to the esoteric wisdom of ancient Egypt that are worth mentioning here.

1. The Emerald Tablet

Attributed to Hermes Trismegistus, the Emerald Tablet is a short, cryptic text that has been highly influential in alchemical and esoteric traditions. It contains the famous phrase "As above, so below," suggesting a correspondence between the macrocosm and microcosm. This idea resonates with Egyptian beliefs in the interconnectedness of the cosmos and humanity, highlighting the pursuit of hidden knowledge.

2. The Coffin Texts

These texts, which evolved from the Pyramid Texts, were inscribed on coffins during the Middle Kingdom period (circa 2040–1782 BCE). They contain spells and teachings aimed at guiding the deceased through the afterlife, revealing insights into the Egyptian understanding of the soul, immortality, and the nature of existence. The emphasis on transformation and rebirth parallels other esoteric traditions.

3. The Pyramid of Pepi II

The Pyramid of Pepi II (circa 2278–2184 BCE) contains various inscriptions and spells that reflect advanced theological and philosophical ideas. Some scholars suggest these texts demonstrate an evolving understanding of the afterlife and divine order, further contributing to our understanding of ancient Egyptian esotericism.

4. The Book of the Heavenly Cow

This mythological text, found in the tomb of the pharaoh Seti I, narrates the story of the goddess Nut transforming into a cow to shelter the sun god Ra. The text explores themes of creation, cosmic order, and the relationship between gods and humanity, embodying the rich symbolism and spiritual depth characteristic of Egyptian wisdom.

5. The Tulli Papyrus

Though controversial and debated regarding its authenticity, the Tulli Papyrus is said to describe an unidentified flying object witnessed over Egypt during the reign of Pharaoh Thutmose III. If genuine, it could suggest that the Egyptians possessed knowledge or experiences that transcended ordinary understanding, aligning with themes of esoteric wisdom and the exploration of the cosmos.

6. The Westcar Papyrus

This papyrus, dating to the Middle Kingdom, contains a series of tales about the wonders performed by ancient Egyptian magicians and the wisdom of the pharaohs. The stories emphasize themes of divine favor, the power of magic, and the relationship between the king and the gods, highlighting an understanding of the supernatural and its role in governance and society.

7. The Book of Amduat

This funerary text, often found in the tombs of pharaohs, describes the journey of the sun god Ra through the under-world each night. It outlines the twelve hours of the night and the trials faced by the sun, representing themes of rebirth, resurrection, and cosmic order. The text is rich in symbolism and esoteric knowledge, illustrating the Egyptians' beliefs about the afterlife and the cyclical nature of existence.

8. The Ritual of the Opening of the Mouth

This ancient ritual, performed during mummification and funerary practices, was believed to animate the deceased, allowing them to eat, speak, and breathe in the afterlife. The rituals involved complex incantations and ceremonies, reflecting deep esoteric knowledge about the nature of life, death, and the soul's journey.

9. The Destruction of Mankind

Found in the Pyramid Texts, this narrative describes a cosmic struggle and the divine intervention necessary to restore order. It embodies themes of chaos, creation, and the maintenance of Ma'at (truth and balance), illustrating the Egyptians' understanding of the universe as a dynamic interplay between chaos and order.

10. The Tomb of Tutankhamun

The artifacts and inscriptions found in the tomb of Tutankhamun offer rich insights into ancient Egyptian beliefs and practices. Items like the Ankh (symbol of life) and the use of protective spells indicate an understanding of the afterlife and the importance of safeguarding the soul's journey, showcasing the wisdom embedded in funerary customs.

11. The Serpent Magick

Various texts and artifacts, including depictions of the Uraeus (the cobra), emphasize the protective and transformative qualities of serpents in Egyptian mythology. The Uraeus represented sovereignty, divine authority, and the protective power of the goddess Wadjet, reflecting a sophisticated understanding of symbols and their esoteric significance.

These discoveries underscore the richness of ancient Egyptian esoteric wisdom, revealing a complex interplay of spiritual ideas that continue to intrigue scholars and seekers alike. The texts and artifacts reflect a profound understanding of the universe, the divine, and the human experience, inviting us to explore the depths of ancient knowledge and its relevance today.

In our quest to decode the language of the gods, we find ourselves at the crossroads of ancient wisdom and modern inquiry, where the hieroglyphs of Egypt whisper secrets of a civilization steeped in profound knowledge. As we explore the Pyramid Texts, with their intricate invocations for the afterlife, and the mystical teachings embedded within the Book of the Dead, we realize that these texts serve as portals to an esoteric understanding of existence.

Could it be that the enigmatic symbols carved into temple walls and the lost scrolls of the Al-Hammadi Library and the Library of Alexandria hold keys to a forgotten legacy of spiritual insight? The possibility that ancient Egyptians were not merely builders of grand monuments but custodians of a universal wisdom — one that may even echo the profound metaphysical concepts found in the Vedas — challenges us to rethink our understanding of history and the interconnectedness of human thought.

In deciphering this divine language, we embark on a journey not only into the past but also into the depths of our own consciousness, seeking to reclaim the lost wisdom that may still guide us today.

Jay's visit to the temple of Horus

As Jay stepped into the grand Temple of Horus, the scorching Egyptian sun seemed to pale in comparison to the awe that filled his heart. He had traveled all the way from India, driven by his passion for ancient cultures and their connections. Being a devout student of the Vedas, Jay had always felt a deep affinity for ancient wisdom. Yet, nothing could have prepared him for what he was about to experience in this sacred space, far from his homeland.

The towering stone columns of the temple were adorned with intricate hieroglyphs, telling the story of gods and pharaohs long gone. As Jay wandered deeper into the inner sanctum, a strange feeling settled over him — a mix of familiarity and something more profound. The inscriptions on the walls, though foreign in appearance, stirred something inside him, as if he had seen them before.

He approached one particular wall, its carvings more faded than the rest, and as his fingers traced the symbols, a word formed in his mind — *Agni*, the Vedic god of fire. His breath caught in his throat. He could hardly believe it, but the feeling was undeniable. The hieroglyphs seemed to convey a concept of cosmic energy, a principle he had studied in the *Rigveda*. Could it be possible that these ancient Egyptians had been speaking of the same universal truths?

Jay's heart raced as he moved along the wall, eyes scanning each symbol, each line. It wasn't just a trick of his mind. The depictions of *Ra*, the sun god, mirrored so closely his understanding of *Surya*, the Vedic sun deity, both revered as sustainers of life. But it wasn't the gods that struck him most — it was the cosmic principles they represented. Life, death, rebirth, and the cyclical nature of existence, concepts that were foundational to his Vedic studies, echoed in the art and scripture before him.

And then he saw it, an inscription that shook him to his core. It spoke of the "Eye of Horus," a powerful symbol in Egyptian mythology, representing protection, healing, and restoration. But as Jay stood there, it wasn't Horus he thought of — it was the all-seeing *Chakshu*, the Vedic concept of divine vision. The parallels were too clear to be a mere coincidence.

A shiver ran down his spine as he whispered to himself, "*It's all connected.*"

He remembered a conversation with his guru back in India, who had once suggested that the ancient cultures were not isolated but were tapping into the same universal truths.

"*The Vedas*," his guru had said, "*are not just for India. They speak of the cosmic laws that govern all creation.*"

Standing there, in the heart of Egypt, Jay felt as though those words were being confirmed in the most unexpected way. As the ancient air of the temple wrapped around him, he realized that the inscriptions on these walls spoke a language older than kingdoms, older than borders. It was the language of the universe, one that transcended civilizations, speaking to those who had the eyes to see and the heart to understand.

Jay left the Temple of Horus that day with a deep sense of wonder. The world was far larger, far more connected than he had ever imagined. His journey to Egypt had turned into something more profound than he had planned — he had come to study the past, but he left feeling as though he had touched the very heart of the eternal.

"Here I am, O Ra (Sun-God), I am your son, I am a soul ... a star of gold ..."

— Pyramid Texts, lines 886–9

5. The Gods Among Us: Ancient Deities and Their Influence

In the heart of the desert, where sands whisper secrets of ages past, the legacy of ancient Egypt continues to captivate our imaginations.

In this chapter, I invite you to explore a world where divine beings walked alongside mortals, shaping destinies and the fabric of society. From Osiris, lord of the afterlife and a symbol of hope and renewal, to Isis, goddess of fertility and motherhood, these deities were not mere symbols but architects of culture, spirituality, and power.

As we embark on this exploration, we confront questions that challenge our understanding of their significance. What do we truly know about the myths surrounding gods like Ra, whose journey across the sky symbolizes life and rebirth? How did the narratives of Osiris and Isis shape ancient Egyptians' views on death and the afterlife? What hidden meanings lie beneath these stories? Could the symbols associated with these gods — like the ankh, representing life, or the scarab, signifying regeneration — hold esoteric knowledge? How did the ancient Egyptians interpret these symbols, and what relevance do they have today?

Our sources of information about Osiris, Ra, Isis, and Anubis in ancient Egypt come from various types of archaeological and textual evidence. Some of the key sources are:

1. **Religious Texts:**

   ♦ **The Pyramid Texts:** These are some of the oldest religious writings, dating back to the Old Kingdom (c. 2686–2181 BCE). They include spells and rituals aimed at guiding the deceased through the afterlife and often reference Osiris, Ra, and Anubis.

   ♦ **The Coffin Texts:** Emerging during the Middle Kingdom (c. 2055–1650 BCE), these texts expanded on the themes found in the Pyramid Texts, containing spells intended to protect the deceased and help them in the afterlife.

   ♦ **The Book of the Dead:** This collection of spells and illustrations served as a guide for the deceased in the afterlife, detailing the journey through the underworld and emphasizing the roles of gods like Osiris and Anubis.

2. **Mythological Literature:**

   ♦ **The Myth of Osiris:** Numerous versions of this myth exist in various texts, including the writings of later authors such as Plutarch in "De Iside et Osiride," which recounts the story of Osiris, Isis, and Horus.

3. **Temple Inscriptions and Reliefs:**

   ♦ Temples dedicated to these gods, such as the Temple of Isis at Philae and the Temple of Karnak dedicated to Amun-Ra, contain inscriptions and wall reliefs that depict various myths and highlight the significance of these deities in ancient Egyptian worship.

4. **Funerary Art and Artifacts:**

   ◆ Artifacts, such as statues, amulets, and other burial items, often depict these gods, providing insights into their iconography and how they were perceived in relation to death and the afterlife.

5. **Archaeological Findings:**

   ◆ Excavations of tombs and burial sites have uncovered various artifacts and inscriptions that relate to the worship of Osiris, Ra, Isis, and Anubis, revealing the practical aspects of ancient Egyptian religious practices.

6. **Historical Accounts:**

   ◆ Writings from ancient historians, such as Herodotus and Diodorus Siculus, provide external perspectives on Egyptian religion and its gods, contributing to our understanding of their significance in ancient society.

These sources collectively offer a multifaceted view of how Osiris, Ra, Isis, and Anubis were worshiped and understood within the context of ancient Egyptian beliefs and practices. However, it is essential to acknowledge that we still do not fully understand all of the source evidence. Many texts and artifacts remain fragmentary or have yet to be deciphered, leaving gaps in our knowledge. Additionally, variations in regional worship and local interpretations of these deities can complicate our understanding of their roles and significance.

The evolving nature of Egyptian religion over millennia also means that the meanings and practices associated with these gods likely changed, further obscuring a complete picture. As scholars continue to study these ancient materials, we remain on a journey of discovery, piecing together the rich fragments of beliefs that shaped ancient Egyptian spirituality.

In unraveling the mysteries of these deities, alternative interpretations emerge, prompting us to rethink their origins and significance. Were these gods manifestations of divine power, or could they have been visitors from outer space, as some theories suggest? What evidence supports such radical ideas? For instance, could the elaborate depictions of these gods represent advanced ancient technology or metaphors for natural phenomena, such as the flooding of the Nile?

Through exploring these provocative questions, we uncover the rich archive of ancient beliefs and challenge historical interpretations. As we seek to understand the ultimate truths hidden within these inspiring myths, we consider what it means to seek the divine in a world steeped in mystery.

Entering the mythology of ancient Egypt, we encounter powerful figures intricately woven into the afterlife. Ra, the radiant sun god, illuminates creation and life, while Osiris embodies resurrection and eternal life. Isis plays a pivotal role in the myths of death and rebirth, and Anubis, the jackal-headed god, guides souls through the underworld.

Yet, as we delve into their significance, alternative theories prompt us to reconsider the nature of these gods. Some propose that Ra and Osiris reflect ancient astronomical knowledge, with Ra symbolizing the sun's life-giving force and Osiris representing seasonal cycles. Others suggest these deities serve as allegories for societal structures, embodying ideals of kingship and governance.

Furthermore, the question of whether these gods were mythological figures or visitors from advanced civilizations lingers. Could the rituals surrounding Anubis and the afterlife have roots in extraterrestrial encounters or lost ancient wisdom? As we explore these layers of meaning, we uncover profound spiritual insights and engage with a broader discourse on the origins and interpretations of these beliefs, inviting us to ponder how ancient wisdom continues to resonate in our understanding of life and death today.

"The gods of ancient Egypt were not distant deities; they were intimately woven into the fabric of daily life. Egyptians believed their gods influenced everything, from the Nile's flooding to the afterlife, embodying both nature and human experiences."

— Geraldine Pinch

5.1 Ra, Osiris, and Isis: Guardians of the Afterlife

Conventional archaeology views Ra, Osiris, and Isis as significant deities within the ancient Egyptian religious system, integral to their understanding of life, death, and the afterlife. However, it is important to note that our comprehension of these deities is still evolving, with many nuances and local variations in worship that remain unexplored. That being said, it is generally accepted that:

Ra, the sun god, was one of the most important deities in ancient Egypt. Often depicted with a falcon head crowned with a solar disk, he represented light, warmth, and growth. Ra was believed to travel across the sky in a solar boat during the day and journey through the underworld at night, symbolizing the cycle of life, death, and rebirth. His daily renewal was viewed as a metaphor for the resurrection of the sun and the promise of life after death.

The worship of Ra was central to the pharaonic ideology, as the king was often considered a manifestation of Ra on Earth. Temples dedicated to Ra, such as the one at Heliopolis, served as key religious centers. Known in ancient Egypt as "Iunu" or "On", Heliopolis was located near the modern-day suburb of Cairo and served as the cult center for the worship of the sun god Ra (or Atum-Ra). The city's name in Greek, "Heliopolis," literally translates to "City of the Sun," emphasizing its significance as a solar deity's spiritual hub.

Although little remains of the ancient city today, Heliopolis once housed grand temples, obelisks, and monuments dedicated to Ra. Its priests were highly influential in shaping Egyptian cosmology, especially the creation myth involving Atum, who was believed to have emerged from the primordial waters to create the world. The city was also closely linked to the development of the ancient Egyptian solar calendar and astronomical knowledge.

Osiris, typically depicted as a green-skinned figure wrapped in white linen, was the god of the afterlife, resurrection, and agriculture. According to myth, he was the first king of Egypt who was murdered by his brother Set but was later resurrected by Isis. This resurrection made Osiris the ruler of the underworld, where he judged the souls of the deceased.

The myth surrounding Osiris emphasized the themes of death and rebirth, serving as a model for the soul's journey after death. His role in the afterlife was crucial, as he was seen as the judge who determined the fate of souls, weighing their hearts against the feather of Ma'at, the goddess of truth. This belief in Osiris would have fostered hope for resurrection and eternal life, influencing burial practices and the development of elaborate tombs.

Isis, depicted as a woman with a throne-shaped crown, was the goddess of magic, fertility, and motherhood. She played a vital role in the myth of Osiris, using her magical abilities to resurrect him after his death. Isis was also associated with healing and protection, making her a central figure in both personal and state worship.

Her nurturing nature made Isis a beloved deity among the people. She was often invoked in funerary texts and rituals, emphasizing her role in protecting the deceased and ensuring safe passage to the afterlife. Her influence extended beyond Egypt, as her worship spread throughout the Greco-Roman world, showcasing her significance in ancient spirituality.

Significance of the Afterlife

The concept of the afterlife was paramount in ancient Egyptian belief systems. Egyptians viewed life as a journey that continued after death, and they believed that a proper burial and adherence to rituals were essential for a success-ful transition to the afterlife. The afterlife was characterized by the belief in a paradise known as the Field of Reeds, where the deceased could enjoy eternal bliss if deemed worthy.

Key Elements of Afterlife Beliefs:

- ◆ **Judgment:** After death, the deceased would undergo judgment by Osiris, who would weigh their heart against the feather of Ma'at. A heart lighter than the feather indicated a life of virtue, allowing passage to the afterlife.

- ◆ **Burial Practices:** The significance of mummification and elaborate tomb construction, including the in-clusion of grave goods and spells from the Book of the Dead, underscored the belief that the deceased needed protection and provisions in the afterlife.

- ◆ **Continuity:** The afterlife represented continuity with the living world. Ancestral veneration was common, and the living were expected to honor the dead through rituals, ensuring that their spirits remained content and could provide blessings in return.

Overall, the beliefs surrounding Ra, Osiris, and Isis, and the afterlife, shaped not only religious practices but also the social and political structures of ancient Egyptian civilization, reflecting a profound relationship between the divine, the natural world, and human existence.

The origins of these ancient Egyptian beliefs in gods like Ra, Osiris, and Isis, as well as their concepts of the afterlife, are deeply rooted in the early development of Egyptian culture and religion, which evolved over thousands of years.

In the early days, during the Predynastic period (around 4000–3100 BCE), various tribal communities began to form along the Nile. The early Egyptians worshiped a multitude of local deities, often associated with natural elements and celestial bodies. As agriculture developed, deities related to fertility, the Nile, and the sun gained prominence, reflecting the importance of these elements for survival.

These developments were not unique to Egypt. Similar patterns of religious growth can be observed in various ancient societies that arose in conjunction with agriculture and environmental factors.

In ancient Mesopotamia, for example, the Sumerians (c. 4500–1900 BCE) worshiped a pantheon of gods linked to natural forces and agriculture. Deities like Inanna (goddess of fertility and war) and Enlil (god of wind and storms) were integral to agricultural success and city life. The reliance on the Tigris and Euphrates rivers for irrigation mirrored the Egyptian relationship with the Nile, fostering beliefs in divine favor and intervention in crop yields and harvest cycles.

Similarly, in the Indus Valley (c. 3300–1300 BCE), early urban societies developed around fertile river systems. While less is known about their specific deities, archaeological evidence suggests reverence for fertility symbols, such as the mother goddess figurines, reflecting the importance of agriculture and communal well-being. The cyclical nature of planting and harvesting likely influenced their spiritual beliefs, much like in Egypt.

In Mesoamerica, civilizations such as the Maya and Aztecs developed rich religious traditions closely tied to agricultural cycles. The maize god was central to their cosmology, reflecting the significance of corn as a staple crop. Rituals, including offerings and sacrifices, were performed to appease the gods and ensure bountiful harvests, echoing the Egyptian focus on divine favor for agricultural success.

The rise of deities associated with fertility, agriculture, and natural phenomena in these societies reflects a logical response to environmental conditions. As communities transitioned from nomadic lifestyles to settled agriculture, their survival increasingly depended on the success of crops and favorable weather. These patterns illustrate how ancient peoples around the world developed spiritual beliefs in response to their environments, creating rich mythologies that sought to explain and influence the world around them.

By the time of the Early Dynastic Period in Egypt (c. 3100–2686 BCE), certain gods began to unify into a more structured pantheon. Ra emerged as the sun god, symbolizing creation and kingship, while Osiris and Isis represented themes of death and rebirth. Osiris, in particular, became associated with agricultural cycles, reflecting the connection between seasonal renewal and resurrection.

Again, Egypt was not unique in this development of pantheons. Similar patterns can be observed in various ancient societies, where environmental and agricultural factors influenced the development of structured belief systems.

In ancient Greece, for instance, the pantheon of gods, including Zeus, Demeter, and Dionysus, emerged around the time of the Mycenaean civilization (c. 1600–1100 BCE). Zeus, as the king of the gods, represented order and authority, while Demeter was central to agricultural fertility and harvests, embodying the seasonal cycles of planting and reaping. The mythology surrounding these deities provided explanations for natural phenomena and reinforced community values related to agriculture and prosperity.

In the Indus Valley, the early roots of Hinduism can be traced back to agricultural societies that revered deities connected to fertility and nature. Gods such as Indra (associated with rain) and various earth goddesses became central figures as agricultural practices developed. The logic behind this pantheon reflects the dependence on monsoons and crop cycles, much like the Egyptian reverence for gods related to the Nile and seasonal renewal.

The Roman pantheon, which evolved from earlier Etruscan and Greek influences, included deities like Jupiter, Ceres, and Bacchus. Jupiter, akin to Zeus, symbolized authority, while Ceres was worshiped as the goddess of agriculture, embodying the importance of grain and harvests. As Rome transitioned from a republic to an empire, the unification of various deities into a cohesive pantheon served to reinforce the central authority of the state and promote cultural unity among diverse populations.

In Norse mythology, the pantheon emerged in response to the harsh agricultural conditions of Scandinavia. Deities like Odin, Thor, and Freyja were associated with war, fertility, and the natural elements. Odin, as the chief god, represented wisdom and the pursuit of knowledge, while Freyja was linked to fertility and the harvest. The structured pantheon reflects the community's efforts to understand and manipulate their environment, fostering resilience against the challenges of their climate.

We could say that the rise of structured pantheons in these societies illustrates a common logic: as agricultural practices intensified and communities became more complex, the need for organized belief systems grew. These pantheons provided frameworks to explain natural phenomena, cultivate social cohesion, and legitimize political power. By personifying natural forces and societal ideals through deities, these cultures created rich mythologies that resonated with their lived experiences, reinforcing their agricultural reliance and fostering a sense of shared identity.

The Power of Myths

The myth of Osiris and Isis, which developed over time, encapsulated key beliefs about life, death, and the afterlife. The myth of Osiris is one of the most significant and enduring stories in ancient Egyptian mythology, encapsulating themes of death, resurrection, and the afterlife. Here's a detailed overview of the myth and its key elements:

The Story of Osiris

1. Osiris's Ascension:

Osiris was originally the god of agriculture and fertility, revered as the first king of Egypt. He taught the people how to farm and cultivate the land, bringing civilization and prosperity. His benevolent reign made him beloved among the people.

2. Jealousy of Set:

Osiris's brother Set, the god of chaos, storms, and the desert, grew envious of Osiris's popularity and power. Determined to usurp the throne, Set devised a cunning plan to eliminate him.

3. The Murder of Osiris:

Set tricked Osiris into a sealed coffin, claiming it was a gift. Once Osiris was inside, Set threw the coffin into the Nile, where it was carried away and Osiris drowned. This act marked the beginning of Osiris's association with death.

4. Isis's Quest:

Isis, Osiris's devoted wife and sister, was devastated by his death. Using her magical abilities, she searched for his body, overcoming various obstacles and challenges. After a long quest, she discovered Osiris's remains in Byblos (modern-day Lebanon), where the coffin had washed ashore.

5. Resurrection:

Upon retrieving Osiris, Isis used her magic to bring him back to life. This resurrection was pivotal, symbolizing the cyclical nature of life and death. However, Osiris could not return to the land of the living. Instead, he became the ruler of the underworld, overseeing the judgment of souls.

6. The Birth of Horus:

After Osiris's resurrection, Isis conceived a son, Horus. To protect him from Set, Isis hid Horus in the marshes of the Nile Delta. Horus would later grow up to challenge Set for the throne, embodying the struggle between order and chaos.

7. Horus's Revenge and Restoration:

When he came of age, Horus confronted Set in a series of battles. Ultimately, with the support of the other gods, Horus defeated Set, restoring order to Egypt and reclaiming the throne that rightfully belonged to Osiris.

The myth of Osiris illustrates the ancient Egyptian belief in resurrection and the continuity of life beyond death. Osiris's transformation from a murdered king to a ruler of the afterlife embodies the hope for eternal life, influencing funerary practices and the importance of proper burial rituals thereafter.

The conflict between Osiris and Set represents the struggle between good and evil, order and chaos. Horus's eventual victory reinforces the idea of divine justice and the restoration of balance in the world. Osiris's death and resurrection symbolize the annual flooding of the Nile and the renewal of life, linking the myth directly to the agricultural practices of ancient Egyptians.

It's no exaggeration to say that the Osiris myth had a profound impact on Egyptian religion and culture. He became a central figure in funerary texts, and his worship spread widely, influencing later religious thought. The themes of resurrection and judgment would resonate through various cultures, shaping their ideas about the afterlife.

The Greeks, for example, adopted aspects of Osiris's narrative and incorporated them into their own mythologies, often depicting Osiris in relation to the cult of Dionysus, emphasizing themes of resurrection and fertility. The well-known myth of Persephone's abduction by Hades and her subsequent seasonal return to the Earth reflects themes of death and renewal also similar to those of Osiris. Just as Osiris's resurrection symbolizes eternal life, Persephone's return signifies the rebirth of nature, connecting agricultural cycles with the afterlife.

It's interesting to note that many other cultures spread across the world held similar myth motifs. How is that possible? Well, historians point to several factors in their attempts to explain the common patterns, although there are alternative explanations, which we will look at in a while. For now, let's see some examples of myths that contain almost identical elements to that of Osiris:

1. Mesopotamian Religions

Dumuzi/Tammuz: In Mesopotamian mythology, the god Dumuzi (also known as Tammuz) undergoes a similar narrative arc, where he dies and is resurrected, symbolizing the seasonal cycles of fertility and agriculture. His myth is tied to the Sumerian agricultural calendar, reflecting themes of death and rebirth analogous to those found in the Osiris myth.

2. Christianity

Jesus Christ: The story of Jesus in Christian theology bears striking similarities to the Osiris myth, particularly in the themes of death, resurrection, and salvation. Jesus's crucifixion and resurrection are seen as a fulfillment of the promise of eternal life, paralleling Osiris's role as the god of the afterlife and resurrection.

3. Zoroastrianism

Ahura Mazda and Angra Mainyu: In Zoroastrian belief, there is a duality between the forces of good (Ahura

Mazda) and evil (Angra Mainyu), with a final judgment determining the fate of souls. This resonates with the Osiris myth's themes of moral judgment and the weighing of the heart against the feather of Ma'at.

4. Indigenous Cultures

Mayan and Aztec Mythology: In Mesoamerican cultures, deities such as Quetzalcoatl and Tlaloc are associated with agricultural cycles and death. Their myths often include themes of sacrifice, resurrection, and the cyclical nature of life, echoing the foundational elements of the Osiris narrative.

5. Hinduism

Vishnu and Shiva: In Hindu mythology, the cycle of life, death, and rebirth (samsara) is central. Deities like Vishnu and Shiva embody concepts of destruction and regeneration, paralleling the Osiris myth in their roles in the cosmic order and renewal.

Mainstream historians and scholars explain the recurring patterns in myths like that of Osiris through several interrelated frameworks and theories. Here are some key explanations:

1. Structuralism

Structuralist theories suggest that myths from different cultures reflect universal human experiences and cognitive structures. Lévi-Strauss argued that myths often address binary oppositions (e.g., life/death, chaos/order) and serve to mediate these contradictions. This perspective posits that recurring themes in myths arise from the common human need to make sense of the world.

2. Cultural Exchange and Diffusion

Historians note that as cultures interacted through trade, conquest, and migration, they often exchanged stories and religious beliefs. This diffusion of ideas can lead to similarities in mythological narratives, as seen in the parallels between the Osiris myth and those of other ancient cultures. Over time, these narratives may adapt to fit local beliefs and practices.

3. Psychological Archetypes

Jungian psychology proposes that certain archetypes (universal symbols and themes) reside in the collective unconscious of humanity. Myths often express these archetypes, such as the hero's journey or the cycle of death and rebirth. The recurring themes in myths across cultures may thus reflect shared psychological experiences and struggles.

4. Societal Needs and Values

Myths often arise in response to the specific needs and values of a society. For example, agricultural societies may develop myths centered on fertility, death, and renewal, reflecting their dependence on seasonal cycles. As communities face similar existential questions about life, death, and morality, similar mythic themes can emerge.

5. Ritual and Practice

The function of myths in religious rituals can also explain their recurring patterns. Myths often underpin rituals that reinforce community beliefs and values. As societies evolve, rituals may shift, but the underlying myths remain, continuing to shape cultural identity and social cohesion.

6. Natural Phenomena and Environmental Context

Myths frequently arise from humanity's attempts to explain natural phenomena. Stories that explain seasonal changes, natural disasters, and cosmic events often recur across cultures. The Osiris myth, for example, reflects the cycles of the Nile and agricultural renewal, a theme seen in many agrarian societies.

7. Historical Context and Memory

Some myths may originate from historical events that are mythologized over time. As societies remember significant figures, conflicts, or transformations, these stories evolve into myths that carry moral or spiritual lessons, reflecting historical continuity and collective memory.

These explanations attempt to demonstrate that recurring patterns in myths are often the result of complex interactions among cultural, psychological, and environmental factors.

On the other hand, there are many alternative explanations for myths, including narratives like that of Osiris, which often involve speculative theories that challenge conventional interpretations. We have already met some of these ideas in previous chapters, but let's have a quick look at them again.

- Advocates of the Ancient Astronauts theory, such as Erich von Däniken, argue that extraterrestrial beings visited Earth in ancient times, influencing human culture, technology, and religion. They propose that many gods and mythological figures, including Osiris, represent these advanced visitors who imparted knowledge to early civilizations. This idea suggests that similarities in myths across cultures are due to shared encounters with extraterrestrial beings.

◆ Theories of lost advanced civilizations, such as Atlantis, often popularized by thinkers like Plato and later Ignatius Donnelly, suggest that highly developed societies existed long before recorded history but vanished due to cataclysmic events. Proponents argue that the myths of later cultures, like those of the Egyptians and Sumerians, may be remnants of the knowledge and wisdom from these lost civilizations.

◆ The idea that ancient myths were inspired by wise sages or mystics can be traced to thinkers like G. I. Gurdjieff and Rudolf Steiner, who emphasized the role of enlightened individuals in imparting spiritual knowledge. These figures may have communicated complex truths about existence through allegorical myths, leading to the creation of stories that convey deeper meanings, such as those found in the Osiris narrative.

◆ Carl Jung proposed the concept of the collective unconscious, suggesting that all humans share a reservoir of memories, experiences, and symbols. In this view, myths are manifestations of this collective consciousness, reflecting universal themes and archetypes. The recurring elements in myths across cultures, including the Osiris story, arise from shared psychological experiences and existential questions inherent to humanity.

◆ Thinkers like Mircea Eliade have explored the connections between myths and natural disasters. They argue that cataclysmic events — such as floods, droughts, or volcanic eruptions — might have inspired stories of death and resurrection, like that of Osiris. Myths may encode communal memories of historical traumas and responses to environmental changes, reflecting humanity's struggle with nature.

- ◆ Scholars such as Joseph Campbell have emphasized the psychological and sociological functions of myths, positing that they address human fears and desires. Myths can reflect societal values and collective anxieties about death, morality, and the unknown. In this framework, the Osiris myth symbolizes not only agricultural cycles but also deeper human concerns about mortality and the afterlife.

- ◆ Thinkers like Mikhail Bakhtin and Victor Turner have explored the idea of intertextuality, suggesting that myths are not unique creations but evolve through cultural exchanges. This theory posits that stories are retold and adapted, incorporating elements from various traditions. The fluidity of storytelling allows myths like that of Osiris to transform over time, reflecting shared cultural narratives.

These alternative explanations for myths invite exploration into the complexities of human belief and history. Is it possible that the ancient gods of Egypt were extraterrestrial travelers, wise sages, or both? There are certainly many references to the gods that are open to interpretation, and much that we still do not know about these divine figures. Such myths, if they were myths at all, helped to shape the ancients' understanding of life, death, and the cosmos, although how can we be sure that they were interpreted correctly?

The Cult of Ra

The Cult of Ra is one of the most significant religious movements in ancient Egyptian history, particularly during the Old Kingdom (c. 2686–2181 BCE). Ra's worship intensified when the pharaohs positioned themselves as embodiments of Ra, establishing a divine kingship that linked the ruler directly to the god. Temples dedicated to Ra, like Heliopolis, reflected the central role of the sun in Egyptian cosmology.

Early worship of Ra began when he initially emerged as a local deity associated with the sun and light. Over time, his significance grew, reflecting the sun's crucial role in sustaining life. By the time of the Old Kingdom, Ra had become a central figure in the Egyptian pantheon, often merging with other deities. He was associated with Atum, the creator god, and later identified with Horus and Osiris, illustrating the interconnectedness of these divine figures.

Before long, the pharaohs of the Old Kingdom promoted the idea of divine kingship, positioning themselves as earthly embodiments of Ra. This belief reinforced the pharaoh's authority and legitimacy, as they were now seen as mediators between the gods and the people. The title "Son of Ra" became common for pharaohs, linking them directly to the sun god. They adorned their crowns with solar symbols, such as the sun disk encircled by a cobra, representing divine protection and power. This imagery also emphasized their role as bearers of Ra's light and maintainers of cosmic order.

The Cult of Ra also involved various rituals, including daily offerings, prayers, and festivals. One significant festival was the "Wepet-Renenutet," celebrating the new year and the inundation of the Nile, which was linked to Ra's regenerative power.

Ra's daily journey across the sky in his solar barque (boat) represented the cycle of life, death, and rebirth. Each night, Ra was believed to descend into the underworld, where he faced challenges, ultimately rising again at dawn. This is a cycle of hope and renewal, which are central themes in Egyptian spirituality. Alternative interpretations of this imagery highlight how the solar boat could represent celestial navigation and the Egyptians' advanced understanding of astronomy. Ra's boat crossing the heavens may be symbolic of the sun's movement through the constellations, with the different stages of his journey reflecting the interplay of time, space, and cosmic forces.

According to Ancient Astronauts theorists, Ra's solar barque might not be a metaphorical or mythological construct but a depiction of advanced spacecraft technology. The ancient Egyptians, like many other early civilizations, could have misinterpreted extraterrestrial craft and their occupants as gods, incorporating them into their religious narratives. Ra, the sun god, may have been a powerful extraterrestrial being who traveled the skies in what the Egyptians described as a "solar boat," symbolizing a spacecraft that moved across the heavens.

In this context, Ra's journey into the underworld, where he battled the serpent Apep, could represent extraterrestrial missions or battles that occurred out of the view of human observers, possibly in the night sky or beyond Earth. The recurring cycle of Ra's disappearance at night and re-emergence at dawn could be interpreted as a daily return of a spacecraft or an advanced energy source, perceived by the Egyptians as the rising sun.

What we can be sure of is that the worship of Ra profoundly influenced Egyptian art, literature, and architecture as well. Sp potent was his persona that temples dedicated to him became elaborate structures adorned with reliefs depicting his power and mythology.

Over time, the Cult of Ra influenced and merged with other religious beliefs, leading to the creation of composite deities, such as Amun-Ra during the New Kingdom. We can see from this development how flexible Egyptian religion was, and its ability to incorporate various elements into a cohesive belief system.

Ra's significance persisted throughout Egyptian history, influencing later religious practices and thought. His myths and attributes continued to be referenced long after the decline of the pharaonic culture and his influence extended beyond ancient Egyptian religion, later religious traditions in various ways.

During the Hellenistic Period, particularly after Alexander the Great's conquest of Egypt, the Greeks began to syncretize their gods with Egyptian ones. Ra was often identified with the Greek sun god Helios. This blending of deities emphasized the universality of solar worship and reflected the Greeks' efforts to integrate into Egyptian culture.

Greek philosophers, such as Plato and Plotinus, were very influenced by Egyptian thought, including the worship of Ra. The concept of the sun as a source of divine knowledge and enlightenment paralleled their ideas about the Good or the One in Platonic philosophy. Equally, Ra's role as a life-giver and illuminator resonated with these philosophical discussions.

In Roman Egypt, the worship of Ra continued, with emperors being associated with solar deities. The Roman emperor Aurelian (reigned 270–275 CE) promoted the worship of Sol Invictus, the "Unconquered Sun," which shared attributes with Ra. This cult emphasized the sun's power and its role in providing stability to the empire.

In certain Gnostic traditions, elements of Ra's mythology influenced the understanding of divine light and creation. The idea of a transcendent god and the emanation of divine beings can be seen as paralleling aspects of Ra's representation as a source of light and life.

The early Christians also adopted solar symbolism, which can be traced back to Ra. The imagery of Christ as the "Light of the World" parallels Ra's attributes as a solar deity. Additionally, the idea of resurrection in Christianity resonates with the themes of renewal found in the Osiris and Ra myths.

During the Renaissance, scholars revived interest in ancient Egyptian thought through Hermeticism. The Emerald Tablet, a foundational text attributed to Hermes Trismegistus (linked to Thoth) we met previously, drew on Egyptian wisdom and often referenced solar symbolism, echoing the qualities of Ra as a source of enlightenment.

In modern esoteric traditions and contemporary spiritual movements, Ra has been adopted into various esoteric and New Age belief systems. His attributes as a solar deity are often invoked in discussions of energy, light, and personal transformation. The emphasis on solar energy and its association with divinity continues to resonate with modern seekers.

As we can see, Ra's influence is evident across a wide spectrum of cultures and religions, illustrating the enduring legacy of this ancient Egyptian deity. From Hellenistic syncretism to the emergence of Gnosticism and modern spiritual movements, Ra's qualities as a source of light, life, and renewal have left an indelible mark on the evolution of religious thought throughout history.

There are many alternative theories that challenge traditional interpretations of who Ra was, if he existed at all. As I mentioned above, Ancient Astronauts theorists argue that Ra may represent an extraterrestrial being who visited Earth in ancient times. They suggest that Ra's attributes as a sun god reflect advanced technology, such as spacecraft, rather than purely divine qualities. Is it possible that the Egyptians may have misinterpreted these beings as gods due to their advanced knowledge? I believe it is something we should consider.

Another interpretation is that the myths surrounding Ra could be a form of encoded knowledge passed down from these ancient astronauts. Proponents of this idea posit that the intricate astronomical and mathematical knowledge reflected in the construction of pyramids and temples suggests that the Egyptians had contact with more advanced civilizations, potentially from other planets. By this token, the worship of Ra symbolizes humanity's interaction with these beings, who taught them about the cosmos.

Another interesting claim is that Ra's association with the sun reflects knowledge of solar technology that was lost over time. This theory suggests that ancient civilizations may have harnessed solar power or had advanced astronomical insights that were not passed down to later generations. Ra, in this context, could symbolize a once-great understanding of the sun's energy and its significance for agriculture and civilization.

In line with this view, the pyramids and other megalithic structures were not merely tombs or religious sites but rather served as power generators or astronomical obser- vatories. They may have been built based on advanced knowledge of geometry and astronomy, possibly imparted by Ra or ancient astronauts. This perspective posits that the alignment of these structures with celestial events indicates a sophisticated understanding of solar and lunar cycles.

It could even be the case that myths about Ra and other deities are historical records of actual events involving ad- vanced beings or civilizations. Would it be too far-fetched to believe that the battles and journeys described in myths may represent encounters with extraterrestrial visitors or the remnants of lost advanced technologies? When we consider this possibility, it opens a door to a whole new take on history and the secrets of the past.

Although lacking mainstream acceptance due to limited empirical evidence, the above theories encourage us to explore into the intersections of mythology, history, and human imagination, inviting us to reconsider how ancient cultures understood the cosmos and their place within it.

The Cult of Isis

One of the most significant religious movements in ancient Egypt, the cult of Isis evolved over centuries, spreading its influence far beyond the Nile. Isis's roots can be traced to early Egyptian religion, where she was initially associated with fertility, motherhood, and magic. Her narrative gained prominence mainly due to her role in the resurrection of Osiris, which symbolized themes of life, death, and rebirth.

The Cult of Isis flourished particularly during the New Kingdom (c. 1550–1070 BCE), when her worship became more widespread. She was seen as a protective deity, embodying the ideals of motherhood and female strength. Temples dedicated to this powerful goddess, such as the one at Philae, became major centers of worship. Various rituals were associated with her, including processions, offerings, and festivals celebrating the mysteries of life and the afterlife. The "Festival of the Coming of the Inundation" celebrated the Nile's flooding, essential for agriculture, linking Isis to fertility and abundance.

Isis was often associated with wisdom and magical practices, believed to possess profound knowledge of healing and protection. This made her a central figure in magical spells and incantations throughout Egyptian history. In fact, it made her one of the most revered deities in ancient Egyptian religion.

She was believed to possess extensive knowledge of herbs and medicinal remedies, making her a crucial figure in ancient healing practices. People would invoke her name for protection against illnesses and to aid in recovery. Numerous spells and rituals dedicated to Isis were even included in ancient medical texts, such as the Ebers Papyrus, with spells often calling upon her to bestow health and alleviate suffering. She had a vital role in both physical and spiritual healing of her ancient worshippers.

The goddess was also frequently invoked in magical spells and incantations because her knowledge was seen as essential for effective magic. Practitioners would call upon her to enhance their rituals, particularly those related to love, fertility, and protection. She uses her magical skills in the myth of Osiris to resurrect her husband after his murder by Set, demonstrating this was a powerful goddess who also protected and healer through magic.

Many ancient Egyptians wore amulets featuring Isis to invoke her protection, illustrating the potency of her abilities to safeguard against harm and ensure health. It's no wonder that she was so adored by everyday citizens, who relied on her enduring presence in their daily lives and spiritual practices. She also shaped the broader religious landscape, influencing rituals, spells, and the lives of countless worshippers throughout history.

As the Hellenistic period unfolded, Isis's worship spread to Greece and Rome. In these cultures, she was syncretized with other deities, such as Demeter and Venus, emphasizing her attributes of fertility and love. The Romans celebrated her mysteries, reflecting her influence on domestic and agricultural practices.

The mystery cult of Isis soon became popular in the Roman Empire, attracting followers from various social classes. Her cult offered personal salvation and a sense of community, similar to the cults of Dionysus and Mithras. The widespread worship of Isis included elaborate rituals and initiation ceremonies, contributing to the rise of mystery religions during this period.

Some scholars suggest that aspects of the Cult of Isis influenced early Christianity, particularly the veneration of the Virgin Mary. The imagery of Isis nursing her son, Horus, bears a great resemblance to depictions of Mary with the infant Jesus. This connection could illustrate how Isis's attributes were absorbed into new religious contexts.

The legacy of the Cult of Isis continued into the medieval period and beyond, influencing various esoteric traditions. Her symbolism of the divine feminine and magical knowledge found resonance in alchemical and mystical texts, contributing to the Western esoteric tradition.

Let's take a closer look at Isis's role and influence in the esoteric domain so we can get a better image of just how influential and longstanding her reputation became.

1. Resonance in Alchemy

In alchemical texts, Isis was often used as a symbol for transformation and regeneration, reflecting her mythological role in resurrecting Osiris. Alchemists, particularly during the medieval and Renaissance periods, drew on her association with magical wisdom to explore processes of physical and spiritual transformation.

Isis was depicted as a maternal figure, which resonated with alchemists' concepts of the divine feminine as a nurturing force in the natural world. Her role in shaping and protecting life paralleled the alchemical quest for creating new life through transmutation.

Hermeticism also borrowed Isis's symbolism. In Hermetic texts, she was sometimes invoked as a goddess of profound wisdom, offering insight into the mysteries of creation. This connection reinforced her legacy as a figure representing both natural and spiritual knowledge.

2. Medieval Mysticism and Gnosticism

In Gnostic traditions, which flourished during the early centuries of Christianity, Isis was often seen as an embodiment of Sophia, the divine feminine principle of wisdom. The Gnostics' focus on hidden knowledge (gnosis) and spiritual enlightenment closely mirrored the mystery rites associated with Isis in ancient Egypt.

Some medieval Christian mystics subtly incorporated Isis's imagery into their writings, interpreting her as a symbol of divine wisdom or as a prefiguration of the Virgin Mary. This blending of pagan and Christian elements was part of the broader synthesis of spiritual traditions that took place during the Middle Ages.

3. Western Esoteric Tradition

During the Renaissance, there was a revival of interest in classical mythology, mysticism, and esoteric philosophy. Thinkers like Marsilio Ficino and Pico della Mirandola explored the links between Hermeticism, alchemy, and ancient religious traditions, including the worship of Isis. They saw in her a symbol of universal wisdom that transcended cultural boundaries, reflecting a broader interest in syncretism — the blending of different religious and philosophical systems.

Isis also played a role in the symbols and rituals of later esoteric movements, such as Rosicrucianism and Freemasonry. In these traditions, she was seen as a figure representing hidden or occult knowledge, a keeper of the mysteries. For instance, the idea of "lifting the veil of Isis" became a metaphor for uncovering the truths of the universe through spiritual enlightenment.

4. Theosophy and Modern Occultism

In the 19th century, figures like Helena Blavatsky, founder of the Theosophical Society, invoked Isis in her works, most notably in *Isis Unveiled* (1877). Blavatsky saw Isis as a representation of ancient wisdom hidden beneath layers of religious dogma, which could be accessed through spiritual and occult practices.

Isis continues to be revered in contemporary pagan and esoteric movements, such as Wicca and other neo-pagan practices. Her association with the moon, magic, and the divine feminine resonates with modern spiritual seekers who emphasize the role of women and nature in religious experience.

The legacy of Isis extended far beyond her origins in ancient Egypt, leaving a profound impact on Western esotericism. Her symbolism of the divine feminine, magical wisdom, and transformative power became integral to the mystical traditions of alchemy, Gnosticism, Hermeticism, and later esoteric movements. Whether through alchemical texts, Gnostic teachings, or modern occult practices, Isis's influence continues to shape spiritual exploration and the pursuit of hidden knowledge.

Once again, those who support an alternative view of history identify Isis as having a connection with ideas about ancient astronauts or lost advanced civilizations. The Ancient Astronauts theory argues Isis may have been one of those extraterrestrial "visitors" who brought advanced knowledge to early civilizations, particularly in the realms of magic, healing, and resurrection.

In this context, the resurrection of Osiris by Isis could be seen as a metaphor for advanced technology, such as cloning or genetic engineering, carried out by alien beings. Her magical abilities and profound wisdom could be interpreted as advanced knowledge far beyond what ancient humans could have possessed on their own.

Those supporting the lost civilisations theory suggest that deities like Isis could be remnants or symbolic representations of a lost advanced civilization, possibly Atlantis or an early, forgotten society. In this narrative, Isis would represent not just a goddess of magic, but a figure embodying the lost technological or spiritual wisdom of this forgotten society. The elaborate rituals and temple systems dedicated to her could be seen as attempts to preserve the knowledge of this ancient civilization.

Alternative thinkers like Carl Jung and Joseph Campbell explore the idea that Isis, along with other goddesses, represents an archetype within the human collective unconscious. From this perspective, Isis may not represent a real historical figure or extraterrestrial, but rather a universal symbol of the Divine Feminine, nurturing power, and the mysteries of life, death, and rebirth that recur in myths across cultures.

This interpretation aligns with more psychological and symbolic explanations, suggesting that figures like Isis emerged from humanity's shared experience, particularly regarding fertility, the natural world, and the mysteries of life and death.

Whether viewed through the lens of ancient astronauts or the remnants of forgotten wisdom, these interpretations seek to explain the seemingly miraculous powers and deep esoteric knowledge attributed to Isis in ways that go beyond conventional historical and archaeological explanations.

Despite the wealth of information we have about Ra, Osiris, and Isis from ancient texts, temples, and artifacts, much still remains clouded in mystery. Their myths, rituals, and symbols invite endless interpretation, leaving us to wonder: were these gods purely mythological constructs created to explain natural phenomena, or is there something more to their stories? Could they represent remnants of a lost advanced civilization or even beings from beyond our world, as some alternative theories suggest? What ancient wisdom did they truly embody, and how much of that knowledge has been lost to time?

As we continue to uncover and analyze the evidence, we are left with tantalizing questions about the true nature of these deities and their place in the greater mysteries of human history.

"Anubis stands as a paradox within the pantheon of Egyptian gods. As a guide to the afterlife and a guardian of the dead, his jackal form evokes both reverence and fear, embodying the mystery of death itself."

— Joyce Tyldesley

5.2 Anubis: Guide to the Underworld

Anubis was the jackal-headed god who played a vital role in ancient Egyptian mythology as the *Guide to the Underworld* and protector of the dead. Associated with mummification and the afterlife, Anubis was a key figure in ensuring that the deceased safely navigated the treacherous journey through the underworld, or Duat, to reach their eternal destination. His presence was both comforting and fearsome, as he oversaw one of the most significant processes in the spiritual life of the Egyptians: the judgment of the soul.

Why was he often depicted as a man with a jackal's head or as a full jackal? The most logical explanation for this association probably arose from the animals' frequent presence around cemeteries, scavenging near graves. The Egyptians, recognizing the importance of protecting the dead, attributed this role to Anubis, transforming the jackal into a sacred symbol of the god's guardianship over the tombs and the deceased.

With Anubis's primary role being to protect the dead and ensure their safe passage into the afterlife, he was invoked in funerary rituals, sometimes depicted standing over mummified bodies. His association with mummification also made him a guardian of the sacred embalming process, ensuring the proper preservation of the body for the soul's journey.

One of Anubis's most famous roles was during the *Weighing of the Heart* ceremony. In the afterlife, the heart of the deceased would be weighed against the feather of Ma'at, the goddess of truth and justice. Anubis was the one who carefully conducted this judgment, overseeing the balance and determining the fate of the soul. If the heart was lighter than the feather, the soul was deemed pure and granted passage into the afterlife. If the heart was heavier, it was devoured by Ammit, the fearsome creature who awaited the condemned souls.

Although Anubis is not as prominent in myths as Ra, Osiris, or Isis, he plays a significant role in stories surrounding death and the afterlife. When he does crop up in these stories, his actions usually reflect his duties in guiding the dead and ensuring their eternal safety.

For example, in the Osiris Myth, after Osiris is killed and dismembered by his brother Set, it is Anubis who assists Isis in gathering the pieces of Osiris's body and preparing him for resurrection. Anubis embalmed and mummified Osiris, ensuring that the god is properly prepared to be resurrected and to rule over the afterlife. This act solidified Anubis's role as the god of embalming and the preserver of the dead.

In the literary tale of Sinuhe, an Egyptian official who flees into exile, Anubis is invoked during Sinuhe's contemplation of death and the afterlife. He prays that Anubis will guide him and ensure that his body is properly prepared and buried when the time comes, reflecting the deep trust the Egyptians placed in Anubis's care of their souls.

Since the ancient Egyptians were obsessed with the afterlife, the role of figures like Anubis was crucial. After all, it was Anubis who influenced whether the dead would enjoy eternal life or face annihilation. Apart from being responsible for the proper mummification of the deceased's body, Anubis also safeguarded the soul's ability to survive and thrive in the afterlife. Without his intervention, the soul might be left to wander or fail to complete its journey.

The figure of Anubis shares parallels with several other deities and figures from ancient cultures that also emphasized the need for a guide to the afterlife. His role as a protector of the dead and mediator between the living and the divine resonates across different mythological traditions, and his influence is seen in later religious and esoteric systems.

In Greek mythology, Hermes took on the role of psychopomp, guiding souls to the afterlife. Like Anubis, Hermes helped the souls of the dead travel safely to the underworld (Hades). While Hermes had a broader role as a messenger of the gods, his function as a psychopomp, guiding souls along their final journey, mirrors Anubis's role in Egyptian tradition. Hermes was also known for his protection of travelers, including those crossing into the afterlife, just as Anubis safeguarded the journey through the Duat.

Another figure from Greek mythology, Charon, the ferryman of the river Styx, also facilitated the transition of souls into the afterlife. Although Charon did not preside over mummification or the judgment of the soul, he was responsible for transporting the dead across the river to Hades, much as Anubis guided souls through the dangerous Duat.

In ancient Mesopotamian religion, Ereshkigal was the goddess of the underworld, responsible for overseeing the dead and the realm of the dead. While Ereshkigal's role differs somewhat from that of Anubis — since she ruled over the dead rather than acting as a guide — both deities had significant roles in ensuring the proper journey of the soul after death. Ereshkigal's counterpart, Nergal, the god of death, also had some guiding and judgment roles.

In Hindu and later Buddhist traditions, Yama is the god of death and the ruler of the afterlife, similar to Osiris's role in ancient Egypt. Yama also acts as a judge of the dead, weighing their deeds and determining their fate in the afterlife. Yama's function in guiding the deceased to their next stage (whether reincarnation or liberation) and his role in the moral judgment of souls parallel both Anubis and Osiris.

Thanatos, the personification of death in Greek mythology, is another figure linked to the role of escorting souls to the afterlife. Though often depicted as more passive than Anubis or Hermes, Thanatos's presence at the time of death suggests a similar guiding influence. Incidentally, the modern Greek word still used today for "death" is "thanatos" (θάνατος).

Anubis also influenced later mystical and esoteric traditions, particularly in the Hellenistic period, when Egyptian religious concepts merged with Greek ideas. The Egyptian Book of the Dead and its focus on the soul's journey, as well as the role of Anubis in judgment and protection, shaped concepts in Gnosticism and Hermeticism. These esoteric traditions adopted Egyptian themes of afterlife protection, the journey of the soul, and divine intermediaries, incorporating them into broader systems of spiritual knowledge and cosmic balance.

Christianity also incorporated the idea of Anubis in its version of death and the afterlife, with Christian theology forming the idea of the final judgment before entering heaven or hell. While Anubis was eventually replaced by angelic figures such as Michael the Archangel, who acts as a guide and protector of souls in some Christian traditions, the fundamental idea of a guiding, protective figure leading souls to their final resting place remains the same.

Furthermore, the figure of Anubis was adopted into the Western esoteric tradition during the Renaissance, when scholars interested in alchemy, astrology, and the mystical sciences looked back to ancient Egypt for inspiration. In Hermeticism, Anubis was sometimes invoked as a protector of sacred knowledge and a guide through the metaphorical underworld of transformation, in the same way that alchemists sought spiritual enlightenment.

In Freemasonry and other secret societies that drew upon ancient esoteric knowledge, Anubis appears as a symbolic figure representing the transition between life and death, or ignorance and knowledge. His role as a guide and protector of the dead mirrors these societies' emphasis on moral living, spiritual development, and the journey toward enlightenment.

We even see Anubis in modern pop culture, particularly in literature, films, and video games where Egyptian mythology plays a role. He is frequently portrayed as a powerful, mysterious figure who acts as a guardian of the dead or a gatekeeper between worlds, reflecting his enduring image as the protector of the dead and a guide through perilous journeys.

Suggestions have been made by those seeking alternative explanations to the nature of characters from ancient Egypt such as Anubis that he could represent an extraterrestrial visitor. His strange animalistic appearance and association with the afterlife might signify knowledge or technology beyond what was available to the ancient Egyptians. To this extent, the rituals of mummification and the journey through the Duat could be interpreted as metaphors for advanced technologies that might have been brought by these ancient visitors. We also find supporters of the lost advanced civilization theory arguing that the Egyptians may have inherited their detailed knowledge of life, death, and the cosmos from an earlier, highly sophisticated culture.

Whether seen as a divine protector or reinterpreted through alternative history theories, Anubis remains one of the most enduring and mysterious figures in ancient mythology.

What was the Underworld?

The underworld, or Duat, was a realm of transition where souls underwent a complex process of judgment, purification, and potential rebirth. The Egyptians believed that life after death was a continuation of earthly existence, but only if the proper rituals were performed and the soul passed judgment. This made the role of gods like Anubis crucial, as they directly influenced whether the dead would enjoy eternal life or face annihilation.

In order to reach the afterlife, the deceased had to take the perilous passage through the underworld. The journey was dangerous, filled with trials, obstacles, and various demonic beings to test the moral character of the soul in question.

The Book of the Dead and the Pyramid Texts contained spells and instructions to help the dead avoid these dangers and successfully pass through the underworld.

The first important step of the journey was mummification after death to preserve the physical body. The Egyptians believed that the soul had several parts, including the ka (life force) and the ba (personality), and these needed a preserved body to return to in order to survive in the after-life. Without proper mummification, the soul would become restless and unable to complete its journey.

- The ka would dwell in the tomb and required offer-ings of food and drink to sustain it.

- The ba, represented as a bird with a human head, was believed to leave the body during the day and return at night, reuniting with the preserved body for regeneration.

The Weighing of the Heart ceremony was the final judgment that determined whether the soul was worthy of eternal life in the Field of Reeds or would be condemned to destruction. For the Egyptians, the heart was the seat of emotions and morality, which is why it was weighed against the feather of Ma'at, the goddess of truth, balance, and cosmic order. Anubis presided over this ceremony, while Thoth, the scribe god, recorded the outcome.

It's important to consider that, for the Egyptians, the afterlife was not an abstract or distant paradise, but a continuation of life on earth. The goal was to reach the Field of Reeds (or Aaru), a divine realm that mirrored the ideal Egyptian coun-tryside, full of fertile fields, rivers, and abundance. In this idealized version of Egypt, where the Nile's waters flowed and crops thrived, one could enjoy the pleasures of earthly life, free from the struggles of death, disease, and hardship.

In this paradise, the dead would live forever in peace, continuing many of the activities they enjoyed in life, such as farming, feasting, and spending time with loved ones. This promise of eternal life was so central to Egyptian religion and daily life that monumental tombs, such as the pyramids and the Valley of the Kings, were built as eternal homes for the deceased. They were filled with items that the dead would need in the afterlife, such as food, clothes, weapons, and treasures.

The elaborate funeral rites and tombs of the pharaohs reflected the belief that the king would not only live forever but continue to serve as a divine intermediary between the gods and the people, even after death.

Regular Egyptians, though not afforded the grandeur of royal tombs, also invested significant resources in ensuring their burial was proper and that they had the necessary provisions to live well in the afterlife.

Through gods like Osiris and Anubis, and their elaborate funerary traditions, the Egyptians sought to ensure that death was not an end but a transition to a new, perfected existence.

It always strikes me as too much of a coincidence that other cultures across the world also had similar notions of an "afterlife", much like the Egyptians. The recurrence of these "myths" invites us to ponder a more mystical explanation — one that suggests the existence of a secret wisdom that has been lost to the sands of time.

Let's take ancient Mesopotamian mythology, for example. The underworld was known as the "Land of No Return" or Irkalla. Unlike the Egyptian concept, the Mesopotamian afterlife was generally bleak and grim. Souls of the dead descended to Irkalla, ruled by the goddess Ereshkigal, where they lived a shadowy existence, regardless of their moral conduct in life. There was no judgment or differentiation based on virtue or sin as in Egyptian mythology, but the idea of a subterranean underworld was common to both.

Later developments in Mesopotamian religion, especially during the Babylonian period, saw more emphasis on the judgment of the dead, likely influenced by Egyptian concepts. The introduction of the idea of the soul's fate being tied to its deeds echoes the Egyptian Weighing of the Heart ceremony, where moral virtue determined access to the afterlife.

The concept of the afterlife also featured very strongly In Greek mythology, which was governed by Hades. He ruled over the realm of the dead, also called Hades, where souls would travel and, depending on their lives, would be sent to different regions. They may pass to the Elysian Fields (Elysium) if heroic or virtuous, while ordinary souls would go to the bucolic Asphodel Meadows. Tartarus was a place of punishment for the wicked or those who had not led a virtuous life. There was also judgment, similar to the Egyptian system, with three judges — Minos, Rhadamanthus, and Aeacus — deciding the fate of the souls.

In Zoroastrianism, one of the earliest monotheistic religions from ancient Persia, the afterlife was highly moralized. After death, souls would be judged at the Chinvat Bridge and, depending on their deeds, would either pass to Heaven (the House of Song) or fall into Hell (the House of Lies). Like the Egyptian judgment before Osiris, Zoroastrianism emphasized a clear moral dimension to the afterlife. The soul's fate was directly tied to the ethical choices made during life, a theme central to both traditions. The concept of a bridge that divides the virtuous from the damned is also somewhat similar to the Egyptian idea of passing through trials in the Duat.

On the other side of the world, the ancient Aztec belief in the afterlife was complex, with different realms based on how a person died. For example, warriors who died in battle went to Tonatiuh Ilhuicac, the house of the sun. In contrast, ordinary people went to Mictlan (the underworld), where they would face a series of trials over four years before reaching their final resting place. The Maya held a belief in Xibalba, a dangerous underworld that souls had to navigate after death, similar to the Egyptian Duat. The soul's journey was filled with obstacles and dangerous trials.

When we look at Hindu belief, we find that the soul undergoes a cycle of reincarnation (Samsara), with the ultimate goal of achieving Moksha, or liberation from the cycle. The quality of each rebirth is determined by karma — the moral consequences of one's actions in life. Although there is no definitive "underworld" like in Egypt, the journey of the soul through multiple lives and toward a final divine state parallels the Egyptian belief in the soul's journey toward eternity.

Could these universal themes of mortality and rebirth be remnants of a profound understanding that transcended individual societies? While shared human experiences, natural cycles, and cultural exchanges undoubtedly play a role, might there also be deeper, hidden connections linking ancient civilizations? This possibility opens the door to the intriguing notion of ancient knowledge, once revered and perhaps even divine, that has slipped into obscurity.

What truths might lie dormant within these myths, waiting to be rediscovered? Are they echoes of a time when humanity held a greater awareness of the mysteries of existence, guided by an enlightened vision that connects us all?

On Becoming God

One of the curious beliefs held by the ancient Egyptians was that they could achieve divine status in the afterlife, particularly through the concept of becoming one with the gods, most notably Osiris. This belief was closely linked to the idea of rebirth and resurrection, which was central to their understanding of the afterlife.

We can go to several sources to find evidence of this belief. The Pyramid Texts, ancient spells inscribed in the burial chambers of pharaohs, describe the deceased's journey to the afterlife and often include affirmations of the pharaoh's divine status. For example, they might state that the king "becomes one with Osiris" or "is raised up to the heavens."

Lines from the Pyramid Texts, such as: *"O King, you are Osiris; you have gone to the West, you have become Osiris,"* signify the belief that the deceased pharaoh could merge with Osiris, embodying the god's essence in the afterlife.

Another inscription reads: *"I have risen as a god; I have been transformed into a divine being,"* emphasizing the transformation of the deceased into a divine state.

The Coffin Texts also expanded upon the themes in the Pyramid Texts and contained spells designed to help ordinary individuals achieve the same resurrection and divine status as the pharaohs, suggesting that anyone could aspire to become divine through proper rituals and moral conduct.

The spell, *"May I come forth as a god; may I be raised as a star,"* reflects the aspiration for the deceased to achieve a divine or celestial status.

Equally, the invocation, *"I am a god; I have become one with the divine,"* illustrates the belief in attaining godhood after death.

In the Book of the Dead, we find spells that could be used as a guide for the deceased navigating the afterlife. Many spells within it reflect the belief that the deceased could join the gods, embodying traits such as wisdom, strength, and divinity, thereby securing their place in the afterlife.

Statements like, *"I am yesterday, I am tomorrow, I am the eternal god,"* reveal a self-identification and reflection on the idea of transcending human limitations and merging with the divine.

Numerous tombs also contain inscriptions and artwork depicting the deceased alongside various deities, often portraying scenes of judgment and resurrection that illustrate the belief in becoming one with the divine after death.

These quotes and references collectively highlight the ancient Egyptians' belief in the potential for individuals, particularly the pharaohs, to achieve divinity in the afterlife, emphasizing the significance of rituals and moral conduct in this transformation.

The notion of becoming one with the gods encapsulates a complex interplay of spirituality, identity, and the quest for immortality. This aspiration reveals much about the ancient Egyptians' understanding of existence and the divine, illustrating their desire not just for an afterlife but for a transformative union with the very forces that governed their world.

At its core, the idea of merging with the divine may speak to a deep-seated yearning for connection beyond the mortal realm. In a society where the gods were seen as integral to the natural order — shaping the cycles of life, death, and rebirth — the aspiration to attain godhood reflects an intrinsic hope for transcendence. This desire was particularly embodied in the narrative of Osiris, whose death and resurrection became a powerful symbol of renewal. By aligning themselves with Osiris, the ancient Egyptians believed they could share in his divine essence, suggesting that mortality was not an endpoint but rather a gateway to a higher state of being.

Curiously, this belief also points to a sophisticated understanding of the self and its potential. The Egyptians saw the human soul as capable of evolving beyond its earthly limitations, hinting at an advanced philosophical perspective on identity and existence. This notion contrasts sharply with many contemporary religious paradigms, where the divine is often perceived as entirely separate from humanity. In ancient Egypt, the boundary between the divine and the mortal was porous, allowing for the possibility that human beings could not only emulate the gods but also embody their qualities.

Furthermore, this aspiration for divinity was deeply tied to the socio-political fabric of Egyptian life. Pharaohs, as self-proclaimed incarnations of the gods, leveraged this belief to legitimize their authority. The cyclical transition from Horus to Osiris reinforced the idea that rulers could ascend to divine status, ensuring the continuity of divine kingship. This merging of political and religious ideologies underscored the centrality of the gods in shaping societal structures and cultural narratives.

Moreover, the concept of becoming one with the gods invites intriguing questions about the nature of spirituality and the afterlife. What does it mean to be divine? Is it the embodiment of wisdom, strength, or compassion? The ancient Egyptians grappled with these profound questions through their rituals, prayers, and artistic expressions, creating a rich tapestry of beliefs that explored the very essence of existence.

In the broader context of human history, the ancient Egyptian belief in merging with the divine resonates with similar themes found in other cultures, where the quest for immortality and divine connection remains a common thread. This shared aspiration raises compelling inquiries about the universal nature of spiritual longing and the potential for ancient civilizations to have engaged with similar mystical insights, perhaps hinting at a lost wisdom that transcends time and geography.

The ancient Egyptian notion of becoming one with the gods also invites intriguing discussions within the realm of alternative theories of history. This aspiration for divine union raises questions about the origins of such beliefs and whether they might be connected to broader, perhaps even cosmic, phenomena that transcend known historical narratives.

Proponents of alternative theories of history suggest that the sophisticated knowledge embodied in Egyptian religious practices — including their understanding of the cosmos, the afterlife, and human transformation — could have been imparted by visitors from other worlds. These beings might have shared insights that were interpreted through the lens of local belief systems, leading to the development of complex spiritual ideologies. In this view, the aspirations of ancient Egyptians to become one with the gods could represent a remnant of a more profound, shared cosmic understanding that has been obscured by the passage of time.

What if the concept of divinity within Egyptian culture reflected a collective memory of a lost civilization that possessed advanced spiritual and scientific knowledge? The Egyptians might have inherited or developed their beliefs from an earlier society that had already explored the relationship between humanity and the divine. The recurring themes of resurrection, transformation, and unity with the gods could be seen as echoes of a once-universal understanding that transcended geographical and cultural boundaries, raising the possibility that these ideas were not unique to Egypt but part of a larger human experience.

The Vedic Influence

I find it intriguing that many of the key concepts surrounding transformation and becoming 'one' with divine beings in ancient Egypt resonate with fundamental ideas to be found in the Vedas. In some esoteric traditions, particularly those within theosophy, it is believed that the Egyptians received esoteric knowledge from older cultures, such as the sages of the Indus Valley or pre-Vedic civilizations. If so, Egyptian priests may have been initiates of ancient wisdom traditions, which originated in the Indian subcontinent and spread westward. The ancient world may have been more connected than historians would have us believe.

It is useful here to make some comparisons between ancient Egypt and Vedic thought, to give us a better idea of how similar the two are. We can state categorically that the ancient Egyptian concept of becoming one with a god after death and the Vedic notion of achieving enlightenment share significant thematic similarities. Both reflect universal human concerns with the soul's transcendence and union with the divine. Here are some key parallels between the two traditions:

1. Union with the Divine

Ancient Egyptian Notion: In ancient Egypt, the soul's ultimate goal after death was to become one with the gods, particularly Osiris, the god of the afterlife, or Ra, the sun god. The Egyptian "Book of the Dead" details the process where the soul, after passing the trials of the Duat, would be judged in the Hall of Ma'at. If found pure, the soul could merge with Osiris and live in the Field of Reeds, a paradisiacal afterlife, achieving a divine status akin to the gods.

Vedic Concept: In Vedic philosophy, the ultimate goal is *moksha*, the liberation of the soul (Atman) from the cycle of *samsara* (birth, death, and rebirth). This liberation involves realizing the soul's unity with Brahman, the ultimate, all-encompassing reality or universal consciousness. Upon achieving enlightenment, the individual no longer perceives the self as separate from the divine but merges with Brahman, attaining eternal peace and oneness with the cosmos.

2. Overcoming Death and Rebirth

Ancient Egyptian Notion: The Egyptians viewed death as a transition, not an end. The ultimate goal was to avoid the "second death," or spiritual annihilation, by living a righteous life and passing the judgment of Osiris. Those who succeeded became *akh*, a glorified, transfigured spirit that could move freely between the mortal and divine realms, living eternally with the gods.

Vedic Concept: In the Vedic tradition, the cycle of *samsara* is seen as a form of bondage, where the soul is trapped in an endless cycle of birth, death, and rebirth. The realization of *moksha* or enlightenment is the escape from this cycle. Like the Egyptian goal of avoiding the second death, *moksha* is the final liberation where the soul is freed from the limitations of earthly existence and merges with the divine consciousness, ending the cycle of reincarnation.

3. Purification and Ethical Living

Ancient Egyptian Notion: To attain a blessed afterlife and become one with the gods, the deceased had to be judged according to the principle of Ma'at, representing truth, justice, and cosmic balance. Only a light, pure heart — unburdened by sin — could enter the afterlife and unite with Osiris.

Vedic Concept: In Vedic tradition, the path to *moksha* involves living according to *dharma*, the moral and cosmic law that governs ethical conduct and righteous living. Through *karma yoga* (the discipline of selfless action), *bhakti yoga* (the discipline of devotion), or *jnana yoga* (the discipline of knowledge), an individual purifies the soul and moves toward enlightenment. This purification process is necessary for the soul to transcend its limited, individual identity and achieve oneness with Brahman.

4. Spiritual Knowledge and Sacred Texts

Ancient Egyptian Notion: Egyptian funerary texts like the "Book of the Dead" or the "Coffin Texts" were designed to guide the soul through the challenges of the afterlife and ensure its successful union with the gods. These texts provided the deceased with sacred knowledge, spells, and rituals necessary for navigating the underworld and attaining eternal life.

Vedic Concept: Similarly, the Vedic scriptures (such as the Vedas, Upanishads, and Bhagavad Gita) contain wisdom about the nature of the soul, the universe, and the path to enlightenment. The Upanishads, in particular, focus on self-realization and the knowledge (*jnana*) necessary for achieving *moksha*. Like the Egyptian texts, these spiritual teachings are meant to guide individuals on their journey toward divine union.

5. Transcendence of the Ego

Ancient Egyptian Notion: In the Egyptian afterlife, the soul was required to relinquish earthly attachments and ego-driven desires to pass through the various stages of judgment and unite with the divine. The Egyptians believed that the soul must align with the eternal order of Ma'at to transcend the limitations of mortal existence and achieve divine status.

Vedic Concept: In Vedic philosophy, enlightenment involves transcending the ego and recognizing the non-dual nature of existence. The ego, or false sense of self, is seen as the main obstacle to realizing the soul's unity with Brahman. Through spiritual practice and self-realization, the individual dissolves the ego, leading to the realization that the self (Atman) is not separate from the divine (Brahman), thus achieving liberation.

6. Cosmic Order and Balance

Ancient Egyptian Notion: The concept of Ma'at was central to Egyptian spirituality, representing the cosmic balance, truth, and justice. Living according to Ma'at ensured harmony between the individual, society, and the cosmos. Only those who upheld Ma'at in life could hope to transcend death and achieve divine union.

Vedic Concept: Similarly, the concept of *rita* in early Vedic tradition referred to the cosmic order that governed the universe. Living in harmony with *rita* through the observance of *dharma* was essential for spiritual progress. Over time, rita evolved into the broader concept of *dharma*, which dictates moral and ethical behavior, aligning the individual with the cosmic order and facilitating the soul's journey toward liberation.

Both the ancient Egyptian and Vedic traditions emphasize the soul's journey toward divine union, whether through overcoming death and rebirth, living a life of moral and ethical purity, or acquiring sacred knowledge. These parallels reflect stunning shared spiritual concerns about the transcendence of the self and the attainment of ultimate unity with the divine.

Ultimately, the ancient Egyptians' desire to merge with the divine raises profound questions about our own understanding of history, spirituality, and human potential. By examining these beliefs through the lens of alternative theories, we open up new avenues for exploration, challenging conventional narratives and inviting a reevaluation of the connections between humanity and the divine.

This exploration not only enriches our understanding of ancient civilizations but also encourages us to reflect on our own spiritual aspirations and the timeless quest for knowledge that continues to resonate through the ages.

A different type of Christian

Duli had always been a curious soul. Raised in a Hindu household in India, he had been sent to a Catholic school, where he spent his formative years learning about Christianity. The hymns, the prayers, the stories of Jesus and the saints were all familiar to him. Yet, even as he immersed himself in Catholic teachings, he remained connected to the rituals and traditions of his Hindu upbringing — the chants, the festivals, the stories of Krishna and Shiva.

Years later, Duli found himself on a trip to Egypt, a place whose history fascinated him. One afternoon, while wandering through the bustling streets of Cairo, he stumbled upon a small, ancient Coptic church. It was different from the grand cathedrals he had seen in Europe, but something about the quiet simplicity drew him in.

As Duli entered, the rich aroma of incense filled the air, and the sound of chanting echoed softly through the stone walls. He felt a sense of reverence that reminded him of the Hindu temples back home, where the air was thick with devotion. Intrigued, he sat quietly at the back, watching as the Coptic Christians prayed, their heads bowed, crossing themselves in a way that felt familiar yet distinct from what he had seen in Catholic Masses.

After the service, Duli struck up a conversation with a local Coptic Christian named Sami. The two walked through the ancient streets, and Duli shared his background — his Catholic schooling, his Hindu roots, and his curiosity about Coptic Christianity. Sami explained that the Copts were one of the oldest Christian communities in the world, tracing their heritage back to Saint Mark, one of Jesus's apostles, who brought Christianity to Egypt.

As Sami spoke, Duli couldn't help but notice how different Coptic Christianity was from the Roman Catholicism he had learned in school. The liturgy was more mystical, the emphasis less on dogma and more on the divine mysteries of God. The use of icons and the focus on saints felt deeper, more interconnected with daily life. The rich use of symbols, the sacredness of water, bread, and oil reminded him of the rituals in Hinduism — where every element had meaning, every action, a connection to the divine.

In particular, when Sami explained the concept of *Theosis*, where Coptic Christians believed in the potential for humans to become united with God, Duli felt a shiver. It echoed so closely the Hindu idea of *Moksha*, where the soul merges with the infinite divine. Both faiths, in their own ways, believed that the soul could transcend the earthly realm and unite with the divine essence, losing its individual ego in the process.

Sami also shared stories of the Coptic monks who retreated to the deserts, living in isolation to seek spiritual enlightenment. Duli was reminded of the Hindu sadhus, renouncing worldly attachments to find the truth within themselves. The more they spoke, the more Duli realized that these two ancient faiths, which seemed so different on the surface, shared a deeper, universal thread. Both embraced the cyclical nature of life, the journey of the soul, and the importance of symbols and rituals to connect with the divine.

Duli also found similarities in their shared emphasis on devotion and surrender. Just as the Copts would venerate the Virgin Mary with prayers and offerings, Duli thought of his own mother, who placed garlands before the statue of Durga, the divine mother in Hinduism. In both faiths, the divine feminine was revered as a powerful force, a protector, and a nurturer.

But what surprised Duli most was the focus on suffering. Sami explained that in Coptic Christianity, much like in Hinduism, suffering was not seen as a punishment but as a path to spiritual growth. Copts believed that their persecution over centuries had brought them closer to God, just as many Hindus viewed the challenges in life as part of their karmic journey, leading them toward eventual liberation.

By the end of their conversation, Duli was overwhelmed. The Catholic teachings he had learned in school suddenly felt distant, almost foreign, compared to this ancient, mystical Christianity that seemed to hold so many echoes of the faith he had grown up with. He realized that, while religious traditions may have their differences, they often share universal truths, all pointing to the same deeper understanding of the human soul's relationship with the divine.

As the sun began to set, casting a golden light on the ancient city, Duli stood outside the Coptic church, his mind racing with new ideas, his heart full of wonder. He had come to Egypt expecting to learn about history and architecture, but he had found something far more profound — a bridge between the faith he was raised in and the new worlds he had encountered.

"O Horus, you are the Great One in the sky. The sky has opened up, and you have come to the sky as the evening star."

— The Pyramid Texts

6. Egypt's Extraterrestrial Connections: From Gods to Visitors

The gods of Egypt were a powerful lot, able to control various aspects of life, nature, and the world. From the building of the Great Pyramids to the meticulous burial rites, the ancient Egyptians were inspired by their beliefs in these divine beings to accomplish amazing feats. Whether through fear, reverence, or both, they strived to live a life of moral integrity and the proper observance of religious practices if they were to secure a favorable existence after death.

These ancient peoples believed in a highly structured pantheon of gods who were responsible for maintaining the cosmic order (Ma'at), with their Pharaoh taking on the role as the divine intermediary. Some of the most prominent gods they revered included:

Horus, the sky god often depicted as a falcon.

Ra (Re), the sun god and king of the gods.

Anubis, the god of mummification and the afterlife.

Isis, the goddess of magic and motherhood.

Osiris, the god of the underworld and resurrection.

Many of the gods were associated with animals, and often depicted with animal heads or even as animals themselves, such as Bastet (a lioness or cat), Sobek (a crocodile), and Thoth (an ibis or baboon).

But who were the gods of Egypt really? Where did they come from and what evidence can we find of their true nature in the ancient Egyptian sites and culture?

Is it really possible (and perhaps, tangible) that the great gods of Egypt were actually extraterrestrial beings who visited our earth thousands of years ago?

No doubt, you will have heard of this idea if you are familiar with the 'Ancient Astronauts theory', which has gained great momentum over the past 60 years or so.

In this chapter, we will explore the captivating idea that ancient Egypt was not an isolated phenomenon but part of a broader narrative involving contact with beings from beyond our world. Could it be that their gods were not mere mythological figures, but members of an advanced extraterrestrial race — visitors from the stars who imparted their knowledge and shaped the course of human history?

This notion challenges everything we think we know about the origins of civilization and invites us to reconsider whether the great achievements of Egypt were inspired by forces far beyond the Earth itself. If this is the case, where can we find the evidence that upturns the history of mankind as we know it? Read on...

"The Earth was visited in antiquity by extraterrestrial beings who contributed to the development of human culture, technology, and religion. The evidence lies in ancient texts, artifacts, and the pyramids, which can only be explained through the influence of these advanced beings."

— Erich von Däniken

6.1 Ancient Astronauts Theory: Were the Gods Ancient Aliens?

The idea that ancient Egyptian gods were actually extraterrestrial beings, or ancient astronauts, is, for now at least, a theory that modern archaeologists do not embrace easily. Proponents of this view interpret various elements of Egyptian art, texts, and mythology in ways that suggest the presence of advanced beings from other worlds in ancient times. This notion ties into several hypotheses, including the idea of *lost civilizations* — such as *Atlantis* or advanced cultures wiped out by cataclysmic events – all of which arouse our curiosity and imagination.

In relation to the origins of the Egyptian gods and the advanced technology this civilization wielded, the examples below offer insights into how the evidence could be interpreted through the lens of those supporting the Ancient Astronauts hypothesis:

1. Depictions of Gods and Advanced Technology

1.1. Horus

The depiction of Horus as a falcon or a man with a falcon's head might be seen as a representation of an extraterrestrial being who appeared in a form that was familiar and symbolic to the ancient Egyptians. Proponents might argue that the falcon-headed god represents a species of advanced beings with a bird-like appearance.

The "Eye of Horus" could also be interpreted as a form of advanced technology or a symbolic artifact from extraterrestrial visitors that provided protection or insight.

1.2. Ra

Ra's depiction with a solar disk and cobra might be interpreted as an advanced spacecraft or a technological device. The solar boat journey described in texts could be seen as a literal spacecraft traveling across the sky, rather than a symbolic journey of the sun.

Indeed, the solar disk could be imagined as a representation of a spacecraft or energy source, and Ra's daily journey through the sky might be viewed as a technological process rather than a mythological narrative.

1.3. Anubis

Anubis's jackal-headed figure could be seen as an extraterrestrial being with a jackal-like appearance or as a being who played a role in guiding human souls through advanced technology.

The embalming process associated with Anubis might be viewed as a form of advanced medical technology provided by extraterrestrials to prepare the deceased for a journey through space or another dimension.

1.4. Isis

Isis's depiction with cow's horns and a throne-shaped headdress might be interpreted as extraterrestrial beings who had advanced knowledge and protection abilities. Her role in the resurrection of Osiris could be seen as evidence of advanced alien technology that enabled her to perform feats beyond human capabilities.

The sistrum (a musical instrument) Isis holds could be interpreted as a technological artifact or communication device from extraterrestrials, representing a way to interact with or harness alien technology.

2. Texts Describing the Gods

2.1. Pyramid Texts

Descriptions of gods like Ra traveling across the sky in a solar boat might be seen as references to extraterrestrial spacecraft. The idea that Ra "ascends" and "descends" could be interpreted as literal space travel. In fact, the journey of Ra might be viewed as a detailed account of a spacecraft's daily journey around Earth, rather than a mythological representation of the sun's movement.

2.2. Coffin Texts

References to gods transforming themselves or being able to shift forms could be seen as evidence of advanced technology that allows for shape-shifting or transformation. The notion of gods being able to transform might be interpreted as alien beings using advanced technology to alter their appearance or their environment.

2.3. Book of the Dead

The various spells and descriptions of gods aiding in the journey through the afterlife might be interpreted as descriptions of extraterrestrial beings assisting humans with advanced knowledge or technology to navigate another dimension or extraterrestrial realms. The guidance of souls to the afterlife or other astral plane would have been done with the aid of advanced navigation technology or knowledge to assist in their journey.

2.4. The Temple of Hathor at Dendera

The Dendera Zodiac, showing celestial alignments and gods, might be interpreted as a map or record of extraterrestrial visitation or advanced astronomical knowledge. The celestial depictions could be viewed as a record of interactions between ancient Egyptians and extraterrestrials, documenting their advanced understanding of the cosmos.

2.5. Biblical References

Passages from both the Old and New Testaments of the Bible are often cited by proponents of the *Ancient Astronauts Theory* as potential references to alien visitors. We find many descriptions of divine encounters, strange phenomena, or otherworldly beings in the texts that might be interpreted as interactions between ancient humans and extraterrestrials. While these interpretations often conflict with traditional theological readings, they are intriguing in their content and symbolism.

1. Ezekiel's Vision of the Flying Chariot (Ezekiel 1:4-28)

One of the most commonly cited passages is from the *Book of Ezekiel*, where the prophet Ezekiel describes a vision of a flying object that many interpret as a spacecraft. Ezekiel sees a "windstorm coming out of the north" and a "great cloud with fire flashing back and forth" (Ezekiel 1:4). Inside the cloud, he observes four living creatures with "the likeness of a man" but with wings and other fantastic features. These creatures are accompanied by wheels that are described as being full of eyes and able to move in any direction.

Ancient Astronauts theorists, like Erich von Däniken, argue that Ezekiel was witnessing an alien spacecraft with its inhabitants — interpreted as the "living creatures" — possibly piloting the craft. The description of the wheels, in particular, has led some to claim that Ezekiel was describing a form of advanced technology, such as a vehicle or a flying machine, far beyond what humans at the time could have understood.

2. Genesis 6:1-4 — The Nephilim

Another frequently referenced passage is in *Genesis 6:1-4*, where the "sons of God" are said to have taken human wives, resulting in offspring known as the *Nephilim*. The passage reads:

> "When human beings began to increase in number on the earth and daughters were born to them, the sons of God saw that the daughters of humans were beautiful, and they married any of them they chose. The Nephilim were on the earth in those days — and also afterward — when the sons of God went to the daughters of humans and had children by them. They were the heroes of old, men of renown" (Genesis 6:1-4, NIV).

Ancient Astronauts theorists argue that the "sons of God" were extraterrestrial beings who mated with humans, producing a hybrid race of giant, powerful beings — the Nephilim. They claim that this is evidence of early genetic manipulation or breeding programs carried out by aliens.

3. The Fiery Chariots of Elijah (2 Kings 2:11)

In *2 Kings 2:11*, the prophet *Elijah* is taken up into heaven in a "chariot of fire":

> "As they were walking along and talking together, suddenly a chariot of fire and horses of fire appeared

and separated the two of them, and Elijah went up to heaven in a whirlwind."

Ancient Astronauts theorists claim that this "chariot of fire" was actually an alien spacecraft. They suggest that the "whirlwind" might have been the result of a propulsion system or some kind of anti-gravity technology. According to this interpretation, Elijah wasn't taken up by God in a symbolic or divine manner but was instead "abducted" by extraterrestrials.

4. The Transfiguration of Jesus (Matthew 17:1-8; Mark 9:2-8; Luke 9:28-36)

The Transfiguration of Jesus, described in the Synoptic Gospels, is another event that Ancient Astronauts theorists interpret as a potential alien encounter. In this story, Jesus takes Peter, James, and John up a mountain, where he is transfigured: his face shines like the sun, and his clothes become dazzling white. Moses and Elijah appear alongside him, and a bright cloud envelops them.

Some theorists argue that the shining light and cloud could represent a spacecraft or some kind of alien presence. The sudden appearance of Moses and Elijah is also seen by some as a possible example of time travel or extraterrestrial beings presenting themselves in a form recognizable to the disciples.

5. The Vision of the Apostle John (Revelation 4-5)

The Book of Revelation contains many strange and symbolic descriptions that have led some to interpret certain passages as evidence of extraterrestrial contact. In Revelation 4, the apostle John describes a vision of a door standing open in heaven and being invited to ascend into the divine realm. He sees a throne surrounded by "four living creatures," each with a different face (lion, ox, man, and eagle) and covered with eyes.

Like Ezekiel's vision, this description has been interpreted by Ancient Astronauts theorists as potentially describing extraterrestrial beings or technology. The creatures, with their strange appearance and the emphasis on their "eyes," are seen as symbolic of advanced, otherworldly entities.

While some passages in the Bible, such as Ezekiel's vision or the description of the Nephilim, may appear mysterious and open to various interpretations, religious tradition understands them as symbolic representations of divine encounters, angelic beings, or spiritual truths. Since neither of these suggestions can be scientifically verified, why should the Ancient Astronauts Theory be specifically targeted as unfeasible? It is, in essence, no more fanciful than ideas about winged angels in white descending to earth or floating in the heavens on clouds next to a bearded figure known as 'God', right?

3. Symbolism and Mythology

3.1. Creation Myths

The myths describing gods emerging from primordial waters and creating the world might be interpreted as accounts of extraterrestrial beings arriving on Earth and initiating human civilization. The act of creation could be seen as a technologically advanced process rather than a divine one. Atum's creation of other gods could be viewed as a description of alien beings using advanced technology to engineer life and structure society.

3.2. Osiris Myth

The resurrection of Osiris and his role as ruler of the afterlife might be seen as evidence of extraterrestrial technology that allowed for resurrection or extended life. The myth could be interpreted as describing advanced medical or technological practices from alien visitors that enables beings to be brought back to life.

Proponents of the Ancient Astronauts theory interpret ancient Egyptian representations and texts through a lens that suggests extraterrestrial involvement. They view the gods not merely as symbolic figures or mythological constructs but as advanced beings with technology and knowledge beyond what was possible for ancient humans.

It's easy to understand why the idea of past civilizations being visited by extraterrestrial beings should have become so popular, ever since Erich von Däniken first published his groundbreaking book, *Chariots of the Gods?* in 1968.

His idea that many ancient structures, texts, and artifacts could be evidence of extraterrestrial influence fired the public's imagination. He also suggested that many myths and religious stories are based on encounters with alien visitors – an idea that taps into the human fascination with the unknown. We still understand very little about the purpose, meaning, and construction of ancient structures, like the Pyramids of Giza, Stonehenge, and the Nazca Lines. The idea that extraterrestrials might be responsible offers an intriguing solution to these legendary mysteries.

Following on from von Däniken's work, the Ancient Astronauts theory took off, with subsequent voices in the field reinforcing the beliefs that extraterrestrial beings have been visiting Earth for thousands of years, influencing human culture, technology, and possibly even our DNA!

According to the theory, beings from other planets brought advanced knowledge and technology to early human civilizations. Proponents argue that what we've traditionally interpreted as gods descending from the sky or divine beings from the heavens might actually be based on real encounters with technologically advanced extraterrestrial visitors. Rather than viewing these ancient stories as purely mythological, Ancient Astronauts theorists suggest that they should be interpreted more literally, as records of interactions with alien beings.

"While many point to Egyptian hieroglyphs and artifacts as evidence of extraterrestrial encounters, it is crucial to approach these interpretations critically. However, the enduring fascination with ancient aliens in Egypt speaks to our desire to understand the mysteries of our past."

— Brian Dunning

6.2 Evidence of Extraterrestrial Encounters

Let's look at greater detail into the claims made above and explore what evidence they rely on to support such claims. In general, most Ancient Astronauts theorists refer to the monumental structures to be found around the world, especially the pyramids, the Sphinx, and other massive stone structures found in Egypt. Alignments of such structures with celestial bodies are interpreted as signs of alien influence.

Unusual symbols, hieroglyphs, and art, or depictions resembling modern technology, are also frequently cited as evidence that the ancient Egyptians had access to advanced technology provided by extraterrestrials.

Here are some key examples:

1. **The Pyramids of Giza**
   Due to its sheer size, precision, and alignment of the pyramids with the stars, particularly the Orion constellation, the Giza Plateau near Cairo is seen by Ancient Astronauts theorists as too advanced for the time. They argue that such feats could only have been achieved with the help of advanced extraterrestrial technology.

2. The Dendera Light

A relief in the Temple of Hathor at Dendera, just 60 kilometers from Luxor, shows a figure holding what some interpret as a large bulb-like object. Inside the object is a snake, connected to a lotus flower at the base, and a cable-like object extending to a box. Some theorists suggest that this image represents an ancient electrical device, often referred to as the "Dendera Light," implying that the Egyptians had access to advanced technology, possibly provided by extraterrestrials.

3. Helicopter Hieroglyphs

At the Temple of Seti I in Abydos, an ancient city in Upper Egypt, a panel of hieroglyphs includes shapes that some believe resemble modern vehicles, such as a helicopter, a tank, and a plane. Believers argue that these hieroglyphs are evidence of knowledge of advanced technology, possibly gained through contact with extraterrestrials who introduced or showed these devices to the ancient Egyptians.

4. The Tulli Papyrus

Allegedly found in the Vatican's archives, this papyrus contains ancient Egyptian text that describes "fiery disks" in the sky. Ancient Astronauts theorists claim this is an account of UFO sightings in ancient Egypt, suggesting that the Egyptians witnessed extraterrestrial spacecraft.

5. The Dogon and Sirius Connection

While not directly Egyptian, this belief is often linked to ancient Egypt due to cultural connections with the Dogon tribe of Mali, who claim descent from ancient Egyptians. The Dogon tribe has detailed knowledge of the Sirius star system, particularly Sirius B, a star invisible to the naked eye and only confirmed by modern astronomy in the 20th century. Some theorists argue that this knowledge was passed down from extraterrestrials who visited ancient Egypt and imparted advanced astronomical information.

6. The Osirion at Abydos

Back in Abydos again, we find the Osirion: a megalithic temple structure believed to predate the surrounding temples by thousands of years. Its construction is more reminiscent of earlier, more primitive, yet massive stonework. Proponents argue that its sophisticated construction techniques and the precision of the massive stone blocks suggest the involvement of advanced technology, possibly from extraterrestrial sources.

7. Depictions of the Sky Disc

Across various tombs and temples, Egyptian art often depicts a winged disk or circular object in the sky, usually associated with the sun god Ra. Some theorists interpret these disks as representations of UFOs or extraterrestrial spacecraft, rather than symbolic depictions of the sun or other celestial phenomena.

8. Fallen Angels

The concept of fallen angels in ancient texts like the Book of Enoch lends itself to the idea that fallen angels could have been ancient astronauts interacting with humans. Rather than spiritual entities, these fallen angels could have been visitors to earth, probably visiting places like Egypt, and later interpreted as celestial entities.

While these interpretations make for interesting reading, it's worth pointing out that they are not accepted by mainstream scholars, who generally attribute these artifacts to religious symbolism, artistic expression, or the advanced but human capabilities of ancient Egyptian civilization. Those who support the ancient Astronauts Theory often find themselves at loggerheads with conventional thinking, and their alternative theories of history are not taken as seriously as they perhaps should be.

We know from experience that, the more we learn, the more 'established' parameters are forced to change. What if historians need to rethink their whole timeline on the history of civilization and consider the possibility that they have been wrong all along? A controversial thought, but not infeasible...

The Key Players

Who are the leading voices of the Ancient Astronauts theory and what do they say about the evidence to be found within the monuments and myths of ancient Egypt? Let's find out below!

1. Erich von Däniken

Perhaps the founding father of the Ancient Astronauts theory, Erich von Däniken's book *Chariots of the Gods?* became a bestseller when first released in 1968. It introduced the idea that many ancient structures, texts, and artifacts could be evidence of extraterrestrial influence. Von Däniken suggested that many myths and religious stories are interpretations of encounters with alien visitors.

He based his theory on the idea that advanced technology and alien influence was used in the construction of Egypt's monumental structures, especially the pyramids and the Sphinx. According to von Däniken, the construction of the Great Pyramid of Giza could not have been accomplished by the technology available to the ancient Egyptians. He suggests that extraterrestrials might have helped in the engineering or even built the pyramids themselve, highlighting the precision of the pyramid's alignment with true north and its massive scale.

He also theorized that the Sphinx might be much older than mainstream Egyptology claims, potentially dating back to 10,000 BCE, and that it could have been built by or inspired by extraterrestrial visitors.

Von Däniken pointed to some ancient Egyptian hieroglyphs, such as those found in the Temple of Seti I at Abydos, which appear to depict modern technology like helicopters, submarines, and flying vehicles. He interprets these as clear evidence that ancient Egyptians had contact with advanced alien technology.

2. Zecharia Sitchin

Zecharia Sitchin, author of *The 12th Planet* (1976), focused on the Sumerian civilization and ancient Mesopotamian texts. He interpreted these texts to suggest that the Anunnaki, a group of gods in Sumerian mythology, were actually extraterrestrials who came to Earth to mine gold. His work is more specific in its claims about the origins of these beings and their impact on human civilization.

Sitchin claims the Sumerian Anunnaki were extraterrestrials from the planet Nibiru who influenced early civilizations, including ancient Egypt. He argues that the Egyptian god Thoth, associated with writing, wisdom, and magic, may have been an extraterrestrial figure connected to the Anunnaki. Sitchin also believes that many ancient gods were, in fact, alien beings who visited Earth.

For Sitchin, the pyramids were not just tombs but potentially advanced mining facilities or landing sites for extraterrestrial spacecraft. He speculated that Egypt was part of a larger network of ancient civilizations established by these alien visitors for their purposes, such as mining resources like gold, which he claimed was needed by the Anunnaki for their home planet.

3. Giorgio A. Tsoukalos

Giorgio A. Tsoukalos is the modern face of the Ancient Astronauts school of thought, producing and commenting in the History Channel series entitled Ancient Aliens. He has brought the theory to a broad audience, often discussing how ancient texts and structures could be interpreted as evidence of alien contact.

Tsoukalos argues that the pyramids, especially the Great Pyramid, were not merely tombs but possibly energy-generating structures or devices for communication with extraterrestrials. He points to the lack of burial evidence inside the pyramids and their precise construction as proof that they served a higher, more technologically advanced purpose.

He has highlighted unusual depictions in Egyptian art, such as what some interpret as "lightbulbs" in the Dendera Temple complex, as evidence of alien technology. According to Tsoukalos, these images suggest that Egyptians had access to advanced technologies provided by extraterrestrial visitors. He has also discussed the idea that obelisks were designed to conduct or channel energy, possibly serving as antennas for communication with visitors from other worlds.

4. Robert K. G. Temple

Robert Temple is known for his work *The Sirius Mystery* (1998), in which he posits that the Dogon tribe of Mali had advanced astronomical knowledge provided by extraterrestrials from the Sirius star system. He links this knowledge to ancient Egypt and other ancient cultures, who may have received this know-how from extraterrestrials. He connects the Egyptian goddess Isis, whose mythology is linked to the star Sirius, to these extraterrestrial visitors.

Temple theorized that beings from Sirius visited Earth thousands of years ago, influencing Egyptian religion, astronomy, and architecture. The knowledge about Sirius' binary star system, which he claims was known to the Dogon before modern astronomy confirmed it, is presented as evidence of ancient contact with aliens.

5. David Hatcher Childress

David Hatcher Childress is an author and publisher who has written extensively on ancient mysteries, lost cities, and the possibility of ancient alien contact in his *The Lost Cities* series. He has also been a frequent contributor to the *Ancient Aliens* television series. In his account, ancient Egypt was part of a global civilization that was influenced or even established by extraterrestrials. He often writes about "lost" or hidden technologies that the ancient Egyptians allegedly possessed. Such advanced techniques were used to cut and move huge stones, such as those in the pyramids and other structures, which could not have been accomplished with the tools available at the time. For him, it is most likely that this was made possible through extraterrestrial intervention.

Childress argues that the ancient Egyptians had sophisticated knowledge of mathematics, astronomy, and engineering, which could have come from extraterrestrial sources. He points to the precise measurements of the pyramids, which seem to encode mathematical constants such as pi and the golden ratio, as potential evidence of alien influence. He also speculates that ancient Egypt might have been a colony of a larger, advanced civilization like Atlantis, which he believes was established by extraterrestrials.

6. Graham Hancock

While not strictly an Ancient Astronauts theorist, Graham Hancock has contributed to the broader discussion of lost advanced civilizations and the possibility that ancient cultures had contact with advanced beings, potentially extraterrestrial. His work, such as the book *Fingerprints of the Gods* (1995), often intersects with the themes explored by Ancient Astronauts theorists.

Hancock, alongside other researchers like Robert Schoch, argues that the Sphinx of Giza shows signs of water erosion, suggesting it is much older than traditional Egyptologists believe — possibly dating back to around 10,000 BCE. He theorizes that this earlier date aligns with a time when a lost advanced civilization, potentially connected to extraterrestrials, could have been present.

He also focuses on the alignment of Egyptian monuments with celestial bodies. For example, the three pyramids of Giza are said to align with the stars of Orion's Belt, which he interprets as evidence of advanced astronomical knowledge. Although he doesn't explicitly state that these were extraterrestrials, he suggests that some external, possibly non-human influence, was at work in Egypt's early history.

Hancock goes even further to say that ancient Egypt may have inherited knowledge from an older, highly advanced civilization that was either destroyed or hidden. While he doesn't directly link this to aliens, his theory intersects with those who believe that extraterrestrials may have played a role in shaping ancient cultures.

More Extreme Extraterrestrial Theories

Several ideas within the broader Ancient Astronauts hypothesis go beyond the already controversial claims mentioned above and delve into even more extreme territory. These ideas often push the boundaries of speculation and have garnered attention for their boldness, but they also face significant skepticism, even within the alternative history community. Here are some of the more extreme ideas:

1. Human-Alien Hybrids

Some proponents believe that ancient extraterrestrials not only visited Earth but also genetically engineered early humans, creating human-alien hybrids. This theory suggests that the gods of ancient mythologies, who often interbred with humans, were actually aliens manipulating human DNA to create a more advanced species.

David Icke, a British conspiracy theorist, has suggested that ancient beings (often described as reptilian) manipulated human genetics to create a hybrid ruling class. This idea has been expanded by others who claim that ancient rulers and even modern elites may carry extraterrestrial DNA.

2. The Moon as an Artificial Satellite

Some extreme Ancient Astronauts theorists propose that the Moon is not a natural celestial body but rather an artificial satellite placed in Earth's orbit by extraterrestrials. They argue that the Moon's precise size and distance from Earth, which allows for perfect solar eclipses, is evidence of its artificial origin.

Authors like Christopher Knight and Alan Butler, in their book *Who Built the Moon?* explore the idea that the Moon was deliberately placed in orbit by an advanced alien civilization to influence life on Earth, including the development of human civilization.

3. Ancient Nuclear Wars

Ancient texts and ruins, such as those in India's Mahabharata or the ruins of Mohenjo-Daro, describe or result from nuclear warfare waged by extraterrestrial beings thousands of years ago. Proponents claim that certain ancient cities show signs of extreme heat or radiation that could only be explained by nuclear explosions.

The idea was popularized by authors like Erich von Däniken and David Hatcher Childress, who point to descriptions of "weapons of mass destruction" in ancient epics as evidence of advanced, extraterrestrial technology used in ancient times.

4. The Reptilian Hypothesis

Reptilian extraterrestrials, often referred to as "Reptilians" or "Lizard People," have been secretly influencing human affairs for millennia. Some believe these beings have taken human form or interbred with humans to create a hidden ruling class that controls world events from behind the scenes.

David Icke is one of the most vocal proponents of this idea, arguing that many world leaders and influential figures are either Reptilians or controlled by them. He claims that these beings have been manipulating humanity since ancient times, often being worshiped as gods.

5. Ancient Stargates and Wormholes

Some theorists suggest that ancient civilizations had access to stargates or wormholes — advanced technological devices that allowed for instantaneous travel across vast distances in space. They argue that structures like the pyramids or the ruins at Baalbek, Lebanon, were built as portals or landing sites for interstellar travel.

Researchers like William Henry have explored the idea of stargates, interpreting ancient myths and texts as references to these advanced technologies. They believe that certain ancient sites were designed to harness or stabilize the energy needed to operate these portals.

6. Ancient Time Travel

A more speculative idea is that ancient civilizations had access to time travel technology, either provided by extraterrestrials or developed on their own. Proponents argue that this could explain some of the advanced knowledge found in ancient cultures or the sudden appearance of certain technologies.

Though less popular, some fringe theorists have speculated about time travelers influencing ancient events or leaving behind evidence in the form of anachronistic artifacts, sometimes called "out-of-place artifacts" (OOPArts).

7. Atlantis as an Alien Colony

The legend of Atlantis, the advanced civilization described by Plato, is interpreted by some as a real alien colony that existed on Earth. Proponents argue that Atlantis was either founded by extraterrestrials or humans with advanced alien technology and destroyed in a cataclysmic event, possibly due to a conflict with other extraterrestrial beings.

Authors like Ignatius Donnelly and later, more extreme theorists, have proposed that Atlantis was a hub of alien-human interaction, with its downfall marking the end of a golden age of extraterrestrial influence on Earth.

8. The Ancient Astronauts War

Some suggest that the gods of ancient mythologies, such as the pantheons of Greece, Egypt, and Mesopotamia, were actually warring factions of extraterrestrials who used Earth as a battleground. These wars were witnessed by humans and later mythologized as stories of gods and heroes.

This idea is often explored in science fiction but has also been suggested by some Ancient Astronauts theorists, who interpret ancient myths and epics as accounts of extra-terrestrial conflicts.

These extreme ideas push the boundaries of the Ancient Astronauts theory by incorporating elements of advanced technology, genetic manipulation, and hidden extraterres-trial agendas. While they are widely regarded as speculative and controversial, they continue to captivate the imaginations of those interested in alternative histories and the possibilities of alien influence on human civilization.

The Akashic Record

I want to talk a little about the Akashic Record, which you may or may not be familiar with. This concept can be found in various spiritual and mystical traditions that refers to a metaphysical compendium or "library" of all knowledge, including all human experiences, thoughts, emotions, and events throughout history. Perhaps it can be equated with the Hall of Records supposedly hidden beneath the Sphinx, which we discussed in the previous Chapter.

The term "Akashic" is derived from the Sanskrit word "Akasha," meaning "ether" or "sky," which in Hindu and Buddhist cosmology refers to a fundamental substance or essence that pervades the universe.

The Akashic Record is not considered a physical place. Instead, it is described as existing in a non-physical plane of existence or a higher dimension beyond the material world. It is often depicted as being part of the collective unconscious or the universal mind, accessible through states of heightened spiritual awareness.

Different spiritual traditions and esoteric teachings may describe the Akashic Records differently, but they generally agree that it is a non-physical, omnipresent "storehouse" of knowledge. Some compare it to the concept of a holographic universe, where every part contains the whole, and thus all information is available everywhere at all times.

The Akashic Record is linked to the Ancient Astronauts theory primarily through the idea that ancient civilizations might have accessed advanced knowledge or technology, possibly through contact with extraterrestrial beings, that was recorded or stored in this metaphysical "library."

The key aspects of the concept are this:

- **Universal Knowledge:** The Akashic Record is believed to contain every action, thought, and experience that has ever occurred, as well as all potential future events. It is often described as a cosmic or spiritual "database" where every soul's journey and all possible futures are stored.

- **Accessing the Records:** In mystical traditions, it is believed that spiritually advanced individuals, such as psychics, mediums, or enlightened beings, can access the Akashic Records. This is typically done through meditation, prayer, or other spiritual practices aimed at connecting with higher consciousness or divine wisdom.

- **Purpose:** The Akashic Records are said to be a source of guidance, wisdom, and insight, helping individuals understand their life purpose, past lives, karma, and spiritual path. Some believe that by accessing the Records, one can gain profound insights into the nature of reality, personal challenges, and the spiritual evolution of humanity.

Historically, the concept of the Akashic Records was popularized in the late 19th and early 20th centuries by Theosophists like Helena Blavatsky and later by the mystic Edgar Cayce. They described it as a spiritual record that could be accessed to reveal hidden truths about the universe and individual destinies. While the Akashic Records are most closely associated with Theosophy, they also resonate with ideas in other spiritual traditions, such as the Hindu concept of karma and the Christian idea of a "Book of Life" that records every person's deeds.

In relation to the Ancient Astronauts theory, it is claimed that extraterrestrial beings visited Earth in the distant past, influencing the development of human civilization by imparting advanced knowledge, technology, and cultural practices. Proponents sometimes suggest that the wisdom these beings shared could be connected to or recorded in the Akashic Records.

The idea is that ancient peoples might have accessed or been given knowledge from these records, enabling them to achieve remarkable feats in architecture, astronomy, medicine, and other fields, which seem inexplicable by the standards of the time.

Some modern supporters of the Ancient Astronauts theory, particularly those in the New Age and esoteric communities, believe that individuals can access the Akashic Records through channeling or other spiritual practices. They argue that some of the knowledge attributed to ancient astronauts could actually be knowledge accessed from the Akashic Records by spiritually advanced individuals or through communication with extraterrestrial entities.

This overlaps with the belief that many religious texts, prophecies, and ancient wisdom might have been influenced by beings who had access to these cosmic records.

The concept of Ancient Astronauts often points to similar myths, symbols, and structures found in disparate ancient cultures as evidence of a common extraterrestrial influence. The Akashic Records are sometimes invoked as a possible explanation for these similarities, with the idea that different cultures might have tapped into the same source of universal knowledge, leading to parallel developments in architecture, religion, and science.

Certain versions of the Ancient Astronauts theory even suggest that extraterrestrial beings are not necessarily physical entities but interdimensional beings who can access higher planes of existence, such as the plane where the Akashic Records are believed to reside. These beings could have shared information from the Akashic Records with early humans, thus shaping their development in ways that appear inexplicable by conventional historical understanding.

The Akashic Records are also linked to the concept of spiritual evolution, where the records contain not only the history of the physical universe but also the spiritual progress of all beings. Some Ancient Astronauts theorists suggest that extraterrestrials have guided human spiritual evolution by helping individuals access the Akashic Records, thus accelerating human development in both material and spiritual realms.

Why Do Some People Believe in the Theory?

That's a good question and one that we can explain to a certain extent.

The idea of Ancient Astronauts provides a cosmic context to human history, suggesting that Earth has been visited by advanced beings from other worlds. This aligns with our broader curiosity about space, other planets, and the possibility of life beyond Earth. Since we don't know much about those topics even today, the theory continues to have its appeal.

Many people are skeptical of mainstream historical and scientific explanations, perceiving them as incomplete or biased, and the belief in Ancient Astronauts proves to be a nice alternative. It challenges conventional narratives and offers another explanation for human history and achievements. Anyone who enjoys exploring such alternative histories and the idea that there's more to the past than what is traditionally taught might easily align with the notion that humanity's origins and accomplishments may be more extraordinary than we thought.

In our search for meaning and purpose, we are drawn to the idea because it fulfills a need. The idea that humanity might have been "chosen" or "guided" by advanced beings can be comforting, offering a sense of significance in a vast, mysterious universe. The Ancient Astronauts theory also resonates with some religious and spiritual ideas, such as the notion of gods descending from the heavens. It can be seen as a modern reinterpretation of ancient myths and religious stories, bridging the gap between science and spirituality.

The idea that we were once visited by extraterrestrial beings often relies on the interpretation of ancient art, texts, and artifacts in ways that suggest advanced technology or knowledge — knowledge that the ancient peoples could not have possessed themselves. For example, depictions of "flying objects" in ancient art or references to "chariots of fire" in religious texts are seen as hard evidence of alien encounters.

And the construction of massive structures with apparent precision, such as the Great Pyramids, is seen by many as proof that ancient astronauts visiting earth assisted in their construction. This additional intriguing layer has become popularized in books and television, through writers like Erich von Däniken, and TV shows such as *Ancient Aliens*. These sources have brought the theory to a wider audience, presenting it in an accessible and entertaining format.

Documentaries, movies, and science fiction that explore the idea of extraterrestrial influence on ancient civilizations have also been influential in shaping our views on the possibilities. Works like the *Stargate* franchise, which fictionalizes the idea of alien beings shaping human history, have played their part in reinforcing the theory's presence in popular culture.

The rise of the internet has also allowed for the formation of communities and forums where people who believe in or are curious about the Ancient Astronauts theory can share ideas, evidence, and theories. This sense of community strengthens the belief system and keeps the idea alive and evolving. Conferences and events dedicated to the discussion of ancient mysteries, extraterrestrial life, and alternative history have created a platform for enthusiasts to gather and discuss their ideas, further popularizing the theory, which inspires people to think creatively about history and the universe.

Wondering about what might be possible feeds on the speculative nature of the Ancient Astronauts idea, which allows for endless possibilities. More exciting and engaging than conventional, often mundane explanations, the notion that extraterrestrials left their mark on history's greatest civilizations definitely appeals to those who distrust institutions, including academic and scientific ones. The Ancient Astronauts theory provides an alternative that bypasses traditional authorities and is particularly appealing to those who feel disillusioned with mainstream explanations.

Questioning the Theory

If the Ancient Astronauts theory is incorrect, what do the critics say to counteract it? Can the advanced knowledge and technology of ancient civilizations be explained rationally, based on scientific investigation? Counterarguments in relation to ancient Egypt challenge the idea of extraterrestrial influence as they provide specific explanations to counter many of the claims.

1. Construction of the Pyramids

Archaeologists and engineers have demonstrated that the ancient Egyptians could have used a combination of levers, ramps, a kind of elevator, and a large labor force to move the massive stones. Evidence of workers' villages near the pyramids, along with tools and infrastructure, supports the idea that the pyramids were the result of skilled labor and meticulous planning.

The alignment of the pyramids with the cardinal points can be explained by the Egyptians' advanced knowledge of astronomy. They had developed methods to observe the stars and use them for navigation and building orientation, as seen in other ancient structures.

2. The Dendera Light

Egyptologists interpret the Dendera relief as a symbolic representation of a creation myth involving the lotus flower (representing the source of life) and a serpent (symbolizing divine energy or power). The "light bulb" shape is seen as a stylized depiction of the lotus flower with a serpent emerging from it, a common motif in Egyptian art.

There is no other archaeological evidence in ancient Egypt that suggests the existence of electrical technology. No traces of electrical devices, wiring, or power sources have been found, which would be expected if such technology were in use.

3. Hieroglyphs and the Helicopter Hypothesis

Egyptologists explain that these "helicopter hieroglyphs" are the result of a palimpsest, where older inscriptions were plastered over and re-carved with new inscriptions. Over time, the plaster eroded, causing the original and later inscriptions to blend together and create shapes that vaguely resemble modern vehicles. This is a well-known phenomenon in ancient inscriptions and has been documented in other contexts.

The surrounding inscriptions and the overall context of the temple's carvings are consistent with traditional Egyptian religious and royal iconography, with no evidence of technological or futuristic themes.

4. Advanced Medical Knowledge

Historians and medical researchers emphasize that Egyptian medical knowledge was based on centuries of empirical observation, experimentation, and the systematic recording of treatments and outcomes. The Ebers Papyrus, for example, reflects the Egyptians' deep understanding of anatomy, gained through mummification practices, and their use of natural remedies derived from local flora.

The advanced medical knowledge of the Egyptians can also be attributed to cultural exchanges with neighboring civilizations. Trade routes facilitated the sharing of medical practices and treatments, contributing to the sophistication of Egyptian medicine.

5. The Pyramids' Alignment and Mathematical Precision

The Egyptians were highly skilled astronomers, who carefully observed the stars for religious and agricultural purposes. The alignment of the pyramids with celestial bodies reflects their religious beliefs and the importance of astronomy in their culture. This alignment was achieved through careful observation and the use of simple tools, such as sighting rods and plumb bobs.

The mathematical precision seen in Egyptian architecture is a result of their developing mathematics, including geometry, which they used in construction and land surveying. The Rhind Mathematical Papyrus, for instance, (found in a tomb in the Theban Necropolis) provides evidence of their understanding of geometry and arithmetic, which were crucial in building the pyramids.

6. Depictions of Gods as Aliens

Again, Egyptologists explain that the depictions of gods with animal heads, such as Anubis (jackal-headed) or Horus (falcon-headed), are symbolic representations of their divine attributes and powers, not literal depictions of alien beings. Animal characteristics were used to convey the god's domain or influence, such as Anubis's association with the afterlife and protection of the dead.

The concept of gods in Egyptian religion is deeply rooted in the natural world and their environment. The use of animal imagery reflects the Egyptians' reverence for nature and their belief in the interconnectedness of all life, rather than any extraterrestrial influence.

Critics of the Ancient Astronauts theory emphasize that the achievements of ancient Egypt can be explained by human ingenuity, cultural development, and the accumulation of knowledge over time. Attributing these accomplishments to extraterrestrial influence would undermine the sophistication and capabilities of ancient civilizations and often relies on misinterpretation or selective use of evidence.

Challenging Historical Interpretations

It's understandable that the Ancient Astronauts theory might be rebuffed by traditional scientific and archaeological communities, given the prevailing reliance on the scientific method. This method emphasizes a structured approach to knowledge acquisition: formulating hypotheses, conducting experiments, collecting and analyzing data, and drawing conclusions based on empirical evidence. It remains the cornerstone of modern scientific inquiry to this day.

However, von Däniken's arguments and those of his cohorts challenge traditional interpretations of ancient history by proposing that mainstream scientists and archaeologists may be overlooking alternative explanations for the anomalies and complexities associated with ancient structures and artifacts. His hypothesis posits that conventional explanations are insufficient for accounting for certain advanced technological and astronomical knowledge evident in ancient civilizations.

Mainstream explanations may not fully account for all the evidence, particularly when it comes to the advanced capabilities demonstrated by ancient civilizations. For example:

◆ Traditional archaeological explanations account for the construction of the pyramids through the use of ramps, levers, and a large workforce. Critics of von Däniken argue that this explanation sufficiently addresses the technological feats of the Egyptians. Yet, Ancient Astronauts theorists suggest that the sheer precision of the pyramids' alignment with celestial bodies, as well as the enormous size and weight of the stones used, might imply extraterrestrial assistance. In that case, such precision and scale challenge the conventional understanding of ancient engineering capabilities.

◆ Ancient Astronauts proponents also suggest that ancient myths, legends, and religious texts might be more than just symbolic or allegorical. They could be historical records of actual extraterrestrial visits. Ancient Sumerian texts, for instance, describe gods who descended from the heavens and imparted advanced knowledge to humanity. Von Däniken and his followers interpret these texts as potential records of contact with extraterrestrial beings rather than purely mythological stories.

The above theories underscore the importance of maintaining an open mind when exploring ancient history. While some ideas may be unconventional, they provoke important questions about our understanding of ancient civilizations.

The possibility of an advanced civilization existing before the last Ice Age, potentially receiving knowledge from extraterrestrials, is an idea that falls within the realm of speculative thought and remains unsupported by concrete evidence. This notion ties into several hypotheses, including the idea of lost civilizations — such as Atlantis or advanced cultures wiped out by cataclysmic events — combined with the Ancient Astronauts Theory.

The Case for a Pre-Ice Age Civilization

One theory that could explain the loss of ancient wisdom, brought to us by extraterrestrials, is the *Pre-Ice Age Civilization Theory* (also known as the *Lost Civilization Theory*). This theory suggests that advanced human civilizations existed long before the traditionally accepted timeline of human history, and that these civilizations were destroyed or lost due to cataclysmic events, such as the end of the last Ice Age (approximately 11,700 years ago).

The problem is that it challenges conventional archaeological and historical understandings, which hold that advanced human civilizations, such as those of Mesopotamia and Egypt, only emerged around 6,000 years ago. Even so, a growing body of evidence has led some to question this timeline. These findings include early megalithic structures, advanced knowledge of astronomy, and discoveries of submerged settlements, all of which suggest that human societies may have been more sophisticated earlier than previously thought.

Proponents of this idea often point to several factors:

1. **Lack of Archeological Evidence Due to Rising Sea Levels:** Since much of human habitation during the Ice Age would have been along coastlines, theorists argue that the melting of glaciers and subsequent sea-level rise could have submerged vast amounts of evidence. For example, submerged cities like *Dwarka* in India and various underwater structures off the coasts of Japan and Cuba fuel speculation that advanced human societies might have existed and left behind no discernible surface traces due to these cataclysmic changes.

2. **Mysterious Megalithic Structures:** Sites like *Göbekli Tepe* (around 9,600 BCE) have pushed back our understanding of early human civilization and led some to speculate that older, lost societies might have existed before recorded history.

The complexity of these structures, built before the advent of agriculture, challenges earlier notions of linear human progress.

3. **Catastrophic Events and Myths of Lost Civilizations:** Many ancient mythologies contain stories of great floods or cataclysmic events that wiped out advanced civilizations. Plato's *Atlantis* is the most famous, described as a technologically advanced society destroyed in a single day and night. The concept has since been tied to theories about extraterrestrial intervention or knowledge.

4. **Advanced but Non-Technological Civilizations:** There is also a possibility that pre-Ice Age societies might have been advanced in terms of social organization, mathematics, or even astronomy, without leaving behind the technological markers we associate with modern civilization. This could align with what we see at sites like Göbekli Tepe, where impressive structures were built by supposedly "primitive" societies.

5. **Memory of Lost Knowledge:** Another speculative theory is that extraterrestrial visitors might have imparted knowledge that was later lost or misinterpreted by subsequent generations. Myths of gods descending from the heavens could be cultural memories of these encounters.

6. **The Younger Dryas Impact Hypothesis:** A more grounded hypothesis that could explain the destruction of any potential advanced civilization is the *Younger Dryas Impact Hypothesis*, which proposes that around 12,800 years ago, a comet or

asteroid struck Earth, causing widespread fires and climatic disruption. While controversial, some researchers argue that this impact may have contributed to the mass extinction of megafauna and human populations, disrupting any developing cultures. The evidence for advanced technology or extraterrestrial involvement may be buried somewhere and perhaps one day we will be advanced enough to discover it!

The construction techniques of monumental structures like Stonehenge and the Great Pyramid of Giza continue to fascinate researchers. Theories about advanced technology or external assistance might provide alternative perspectives on how these achievements were realized. As technology advances, new methods of investigation such as ground-penetrating radar, satellite imagery, and other non-invasive techniques could offer new insights into ancient mysteries.

Future discoveries may one day validate or refute the tantalizing claims of Ancient Astronauts theorists, but until that time, we are left to wonder. Could it be that the ancient Egyptians, with their profound achievements in architecture, mathematics, and astronomy, reached such heights solely through their own cultural and intellectual prowess? Their pyramids, aligned with celestial bodies with stunning accuracy, their knowledge of advanced geometry, and the timeless mystique of their religious practices all point to a society far ahead of its time. But are we to believe that such complexity arose in isolation, with no external catalyst?

Or, were they perhaps the recipients of advanced knowledge — gifts from visitors not of this world, whose influence has been woven into the very fabric of their society? The tantalizing mystery of how such precision and ingenuity could have been achieved by a society seemingly lacking the necessary tools lingers in the shadows of historical debate. Legends of divine beings descending from the heavens, and ancient texts that allude to otherworldly entities, leave room for speculation that extraterrestrial intervention might have played a role in shaping not just Egypt, but other ancient civilizations.

This possibility opens the door to deeper questions: Was the construction of monumental sites like the pyramids and the Sphinx part of a greater, more enigmatic mission? Could the secrets of lost civilizations, erased from the annals of history by natural disasters or cataclysms, hold the key to uncovering the true origin of these ancient accomplishments? Are the stories of gods and demigods that permeate ancient myths in reality depictions of visitors from distant planets, whose advanced technologies were misunderstood by early humans?

As we continue to explore the ruins of our ancestors, decipher ancient texts, and unlock the mysteries of forgotten civilizations, the boundaries between human history and the unknown seem increasingly blurred. For every answer we find, new questions emerge, leaving us to wonder if we are, perhaps, looking at the ruins of not just human endeavor, but of a cosmic collaboration that transcends both time and space.

Some mysteries remain and many aspects of ancient Egypt still appear to be inexplicable at first glance. The deeper we look, the more intrigued we become.

That is the secret of its timeless allure!

Alien Visitor and the Healer in Phoenix

I have always been intrigued by the study of energy medicine and healing modalities. My personal experiences with it have solidified for me the understanding of the bi-directional communication and interdependence between our physical bodies and energetic bodies. Liberating at many levels as this has been, I have always been very curious about the origins of some energetic blockages a healer is sometimes able to see in a person's aura that may come from past life. These are typically the most difficult blockages to heal. Once on this path, a medical doctor I highly respect introduced me to a 'channeler' who spoke in 'star language' and performed 'healings channeled from the great white ones', whom she described as 'alien beings'.

Always the curious explorer, I immediately signed up for a healing session, which was conducted via Zoom. I was certainly intrigued by what followed and I have to admit it was quite an interesting experience. When we had finished, she suggested I do a 'hypnotherapy' session with her to solve a health issue I was then facing. While I very much respect the field of Hypnotherapy, I was not sure I wanted to follow through, but the doctor insisted that I do.

Three weeks later, I found myself flying to Phoenix, checking in a hotel, and waiting for the healer to arrive for my hypnotherapy. She had suggested that I stay there overnight as I may not be in a '100% aware' state of mind right after the session.

We began and, soon enough, I found myself regressing into a past life where I saw the Egyptian pyramids being constructed. The scene looked very different from what is depicted in movies and books as there were some clear machines in operation. It was a couple of minutes before I realized that I was actually observing this from above — like from up in the sky. How could that be? This brought me to look around and I saw what looked like an aircraft standing stationary, upon which I was standing on the viewing deck. I could also see other aircraft, with people wearing what in today's world we would call 'space suits'. I was really intrigued and soon saw much more than I am allowed to write about, but let us just say "the picture" came together for me.

The session lasted for about ninety minutes and after that, I related to the healer what I had seen. She simply listened and smiled. I felt very drained and drank some water on her advice before going to sleep. On waking up the next morning, the memory of this experience became even clearer. I asked myself, *"What did I actually see? Alien visitors & technology in Egypt? Was I traveling with them or, more worryingly, was I one of them?"*

I also pondered a more simplistic explanation: *I was just hallucinating in this dream-like state and my mind was playing tricks on me.*

Years later, I went back to Egypt. When I landed, I felt like the ground kissed my feet. I knew instantly where the important energetically-significant spots were — away from tourist attractions. I worked my way to a pyramid which is neither popular nor very well known, far away from Giza, and I recognized it immediately. This is the one I had seen under construction! I had arrived there on the night of a full moon night, a time when I normally lead a group meditation. On this occasion, I decided to do a solo meditation, delegating the group to a colleague. What a special experience that was.

Let me close by saying that my learning from that meditation was this:

The *ancient past lives within all of us, as learnings and memories we can awaken when we need them for our own growth or in the service of a larger purpose.*

"So vast and awe-inspiring was the legacy of Egypt that her monuments, her wisdom, and her gods have never ceased to fascinate and inspire the world."

— Toby Wilkinson

7. The Legacy of Ancient Egypt: Impact on Modern Society

I would like to begin this chapter by asking the following question:

Why are we still so fascinated by a civilization that first rose from the sands over 5,000 years ago, one that continues to evoke awe and wonder in us all even today?

There is a mystique to ancient Egypt that we cannot fully comprehend —a feeling we get when we stand in front of their monumental achievements and learn of their rich spiritual beliefs. A sense of wonder flows through us as we reflect on their timeless quest for immortality, reaching deep down into our very souls.

While we do not know if they achieved the immortality they so desired, the ancient Egyptians have been immortalized in our hearts and minds. Their knowledge, wisdom, and enigmatic culture are embedded in our collective psyche.

In essence, we are so fascinated by ancient Egypt because it represents a civilization that achieved the extraordinary — both materially and spiritually — while remaining shrouded in mystery. It speaks to our deep human desires for knowledge, power, and immortality, making Egypt's distant past feel intimately connected to our present and future.

The sheer scale and grandeur of Egypt's architectural feats, such as the Pyramids of Giza, the Sphinx, and the temples of Karnak and Luxor, provoke awe. These structures, built thousands of years ago, still stand as testaments to Egypt's advanced engineering and astronomical knowledge. The precision with which they aligned these monuments with celestial bodies, and their mastery over stone, continue to inspire wonder about how these ancient people accomplished such marvels without modern technology.

Their spiritual and mythological world, with gods like Ra, Osiris, and Isis, still captivates our imagination. Their beliefs in the afterlife, the soul's journey, and the importance of eternal life are not only fascinating but have influenced other spiritual traditions. The "Book of the Dead," mummification practices, and their elaborate tombs — filled with treasures for the next life — evoke a deep mystery about their views on existence and mortality, which resonate with humanity's own quest to understand life and death.

Despite advances in archaeology and Egyptology, many aspects of ancient Egypt remain shrouded in mystery. How exactly were the pyramids built? What lies undiscovered beneath the sands of Egypt? What deeper meanings are encoded in the hieroglyphs? These unanswered questions fuel our fascination, as we seek to unlock the secrets of a long-lost civilization that achieved so much yet left us with so much to learn.

The influence of ancient Egypt extends into art, architecture, religion, and modern culture. From the Greco-Roman world to contemporary films, literature, and even modern architecture, Egyptian symbols, aesthetics, and ideas have persisted through millennia. The image of the pharaohs, the allure of hieroglyphs, and the mystical appeal of Egyptian artifacts continue to evoke a sense of timelessness, connecting us to a distant yet profoundly influential past.

Egyptians' obsession with the afterlife and their belief in the possibility of eternal life still resonate with us today. Their incredible efforts to preserve their dead, most famously in the form of mummies, reflect humanity's own enduring quest to transcend death. This cultural preoccupation with the afterlife fascinates us, as it reflects universal human fears and desires, mirroring our own enduring concerns about death, memory, and immortality.

Figures like Tutankhamun, Cleopatra, and Ramses II still fascinate us. The discovery of Tutankhamun's tomb in 1922 sparked a wave of Egyptomania, and Cleopatra's political and romantic dramas with Julius Caesar and Mark Antony have inspired countless works of art and literature. The power, wealth, and mystique surrounding these rulers create a sense of grandeur and intrigue that continues to excite us.

It's fair to say that the legacy of ancient Egypt endures as one of the most remarkable and influential civilizations in human history. The allure of its wisdom, mystery, and innovations remains a testament to its enduring influence on both the ancient and modern world.

In this chapter, we will take a deeper look at the influence of ancient Egypt through the millennia and explore its cultural impact. We will examine its architectural and cultural legacy, as well as consider its relevance today. Not only that: we will reflect on the lessons to be learned from ancient Egyptian wisdom and consider its relevance in our modern world. What can we take from this once thriving civilization to help us navigate world conflict, social divide, and our own existential questions? Does ancient Egypt hold the key to our inherent yearning for truth, harmony, inner peace and enlightenment?

Egypt's Cultural Impact

Ancient Egypt's culture has profoundly influenced modern society, leaving an enduring mark across art, architecture, fashion, film, and even spirituality. Here are some key examples of modern cultural themes that come straight from the land of the gods:

1. Architecture and Design

The influence of ancient Egyptian aesthetics on modern architecture, particularly during the Art Deco movement of the 1920s and 1930s, is a fascinating example of how historical discoveries can shape cultural and artistic trends. The discovery of King Tutankhamun's tomb in 1922 by British archaeologist Howard Carter set off a global wave of "Egyptomania" — an intense fascination with all things Egyptian — that permeated many aspects of Western art, design, and architecture. One good example of Egypt's influence on the Art Deco movement is the Chrysler Building In New York.

Completed in 1930 and designed by architect William Van Alen, the Chrysler Building embodies many of the Art Deco movement's Egyptian-inspired aesthetics. While primarily a celebration of modern industry and the automobile, the building also subtly incorporates Egyptian motifs and geometric patterns in its design. Its iconic spire features a stepped design that recalls the form of ancient Egyptian step pyramids, particularly those from the early dynastic period. The notion of tiered structures symbolized elevation toward the divine in ancient Egypt, and in the Chrysler Building, this is reinterpreted as a vertical ascent into modernity and the future.

The sunburst pattern found in the Chrysler Building's terraced crown and ornamentation is another key feature that echoes Egyptian design. Reminiscent of the Egyptian solar deity Ra, the sunburst symbolizes vitality and power. In Art Deco architecture, the sunburst was often used as a motif to represent progress, energy, and the dawning of a new technological era.

The use of stainless steel in the spire of the Chrysler Building not only represents the mechanization and industrial optimism of the age but also evokes the luxurious materials (such as gold and precious metals) that adorned Egyptian artifacts and tombs. The radiating chevrons and triangular shapes in the spire recall the pointed geometry seen in Egyptian temples and pyramids.

Of course, the Chrysler Building is not the only structure to incorporate Egyptian influences during the Art Deco period. Several other buildings and monuments also reflect this trend, such as the Los Angeles Public Library with its pyramid-shaped roofs, Egyptian-style sculptures, and stylized sunbursts. The Grauman's Egyptian Theatre in Hollywood is another prime example of how the Egyptomania phenomenon impacted entertainment architecture. The theater's façade mimics the look of an Egyptian temple, complete with columns, hieroglyphs, and stylized representations of Egyptian gods.

In addition, many private and public mausoleums built in the 1920s and 1930s often adopted Egyptian motifs, using pyramid structures or obelisk-shaped markers to convey eternal life and power, drawing on the ancient Egyptians' association with the afterlife.

The lotus flower was a key symbol in ancient Egyptian art, representing rebirth and purity, as it was linked to the cyclical flooding of the Nile and the sun's daily rebirth. This motif, along with papyrus plants, frequently appeared in Art Deco designs, either as decorative flourishes in interiors or incorporated into the façades of buildings. Egyptian art, known for its highly stylized, repetitive, and symmetrical geometric designs, also lent itself well to the modernist principles of Art Deco. The use of zig zags, sunbursts, and stepped patterns mimicked the hieroglyphic lines and the architectural details of Egyptian temples, pyramids, and tombs.

The use of Egyptian motifs in Art Deco was not just about aesthetics. Ancient Egypt was perceived as a land of mystery, timeless wisdom, and immense power. To Western designers, Egypt represented a kind of exotic "other," which could lend modern architecture a sense of both gravitas and opulence. In a rapidly industrializing world, Egyptian symbols offered a connection to antiquity, conveying durability and permanence — qualities that architects of the 20th century sought to express in their designs.

Egyptian imagery was also tied to ideas of death and immortality, which resonated with the post-World War I generation that experienced widespread trauma and loss. The Egyptian obsession with the afterlife and monumental tombs (like the pyramids) appealed to an audience looking for stability, continuity, and meaning amid societal upheaval.

We can also point to a more recent architectural reference to ancient Egypt – the glass structure known as the Louvre Pyramid (or Pyramide du Louvre in French). This stunning landmark serves as the main entrance to the Louvre Museum in Paris. Designed by Chinese-American architect I. M. Pei and inaugurated in 1989, its striking design has become an iconic part of the Louvre's architectural landscape, blending the museum's historical grandeur with modernist aesthetics.

The pyramid is an ancient and universal symbol, often associated with strength, stability, and timelessness. In many ways, I. M. Pei's design was inspired by the grandeur and geometric precision of the ancient Egyptian pyramids. Given that the Louvre houses one of the world's most significant collections of Egyptian antiquities, the pyramid design creates a subtle homage to that part of its collection, linking the past to the present.

2. Cinema and Pop Culture

Hollywood's fascination with ancient Egypt has profoundly shaped how Western culture perceives and interacts with this ancient civilization. Egyptian themes — whether grounded in historical events, myth, or legend — have provided a rich source of dramatic, visual, and mystical inspiration for filmmakers over the decades. These representations have not only entertained but also influenced our perception of ancient Egypt, often romanticizing or sensationalizing its culture and history.

Hollywood's early depictions of ancient Egypt set the tone for how the civilization would be viewed in Western popular culture. Films like "The Ten Commandments" (1956), directed by Cecil B. DeMille, presented a Biblical narrative set against the backdrop of an opulent and powerful Egyptian empire. The movie, a massive box-office success, highlighted Egypt's grandeur and wealth, portraying the pharaoh's

court as filled with elaborate costumes, towering statues, and monumental architecture.

While the movie is ostensibly about Moses leading the Israelites out of Egypt, it casts ancient Egypt as a place of extreme power and luxury — often emphasizing the stark contrast between the wealth of the pharaohs and the subjugation of the Israelites. The sweeping landscapes, dramatic score, and larger-than-life portrayal of Rameses II reinforced the idea of Egypt as both awe-inspiring and oppressive.

Similarly, "Cleopatra" (1963), starring Elizabeth Taylor, presented a highly romanticized version of the life of Cleopatra VII, the last active ruler of the Ptolemaic Kingdom of Egypt. The film portrays Cleopatra as both a seductress and a shrewd political figure, focusing more on her relationships with Julius Caesar and Mark Antony than on her role as a leader. The film's lavish production — extravagant costumes, intricate sets, and expansive battle sequences — solidified the association between ancient Egypt and Hollywood excess.

It's fair to say that Hollywood's portrayal of Egypt during this time was often filtered through a lens of Orientalism, which refers to the way Western cultures have historically represented the East — particularly the Middle East and North Africa — as exotic, mysterious, and often backward. In these films, Egypt is portrayed as a mystical and decadent land, filled with intrigue, romance, and danger. The Egyptians are often depicted in either highly idealized or villainous terms, reinforcing stereotypes of Eastern despotism and sensuality.

This romanticization of Egypt through an Orientalist lens not only exoticized the civilization but also distanced it from its historical realities, casting it as a fantasy landscape for Western audiences to consume. For instance, Cleopatra, while a real historical figure, is portrayed in Hollywood as a larger-than-life character, often emphasizing her sexuality over her political acumen, thus reducing her legacy to that of a temptress whose fate was entwined with Rome.

The 1999 film "The Mummy", directed by Stephen Sommers, is an excellent example of how Hollywood continues to draw on ancient Egypt for its blend of horror, fantasy, and adventure. The film revitalized Egypt's mystical appeal in the late 20th and early 21st centuries by using themes of resurrection, curses, and supernatural power, which had long been part of Egypt's cultural narrative in the Western imagination.

"The Mummy" builds on the myth of Imhotep, a real historical figure who was an architect and high priest in the Old Kingdom of Egypt, but who, in the film, is transformed into a cursed figure of supernatural terror. The movie plays heavily on the concept of the curse of the pharaohs, a theme that has been part of Egyptology's pop culture narrative since the discovery of Tutankhamun's tomb in 1922. Supposedly responsible for the deaths of several people associated with the tomb's excavation, the curse has long fueled Western fascination with Egypt as a place of hidden danger and forbidden knowledge.

Hollywood's use of ancient Egyptian themes has had a reciprocal effect on public interest in Egyptology, the academic study of ancient Egyptian history, language, literature, and culture. Even though Hollywood often ignored historical facts, it has nonetheless contributed to increased funding, attention, and participation in archaeological expeditions and museum exhibitions. For example, the 1970s touring exhibition of Tutankhamun's treasures — following the media's long-standing fascination with "the boy king" — was attended by millions, proving the enduring allure of ancient Egypt, much of it shaped by its depiction in cinema.

The legacy of ancient Egypt in modern cinema continues with films like "Gods of Egypt" (2016), which blends Egyptian mythology with fantasy action, or Ridley Scott's "Exodus: Gods and Kings" (2014), which retells the story of Moses in an epic fashion. Despite these films' lack of critical success, they reveal a sustained Hollywood interest in Egypt's rich mythological and historical heritage.

Even in video games, like "Assassin's Creed: Origins" (2017), which is set in ancient Egypt, the civilization is portrayed with a blend of historical accuracy and mythological embellishments. These modern depictions continue to fuel the public's imagination and keep Egypt a prominent fixture in global popular culture.

3. Fashion

While fashion may be seen as a trivial matter by some, there is no doubt that the influence of ancient Egypt on modern fashion has been profound, particularly through recurring waves of Egyptian Revival trends that often coincided with key archaeological discoveries. These trends saw the adaptation of ancient Egyptian motifs, symbols, and styles into contemporary fashion, drawing on Egypt's rich cultural history to inspire both haute couture and mainstream fashion movements. The discovery of King Tutankhamun's tomb in 1922, often referred to as the catalyst for "Tutmania", significantly fueled interest in Egypt's aesthetic, which reverberated across the fashion world and continues to influence designers today.

Fashion during the early 20th century began to embrace the geometric, ornate, and symbolic aesthetics of ancient Egypt, featuring distinct elements like Egyptian collars, scarab motifs, and winged sun disks.

As fashion evolved, ancient Egypt continued to inspire high fashion, particularly in haute couture collections. Leading designers of the 20th and 21st centuries have incorporated Egyptian iconography and aesthetics into their designs, blending historical motifs with contemporary style to create innovative collections that pay homage to this ancient civilization. Christian Dior, for example, has referenced ancient Egypt in several collections. Most notably, in 1947, he created a range of dresses and gowns with intricate beading and ornamentation reminiscent of Egyptian jewelry. Dior's fascination with Egypt was part of a broader movement in post-war fashion, where designers looked to historical styles for inspiration and grandeur. The luxurious use of gold, turquoise, and lapis lazuli in his designs echoed the colors and materials found in Egyptian artifacts.

The avant-garde designer Jean-Paul Gaultier has frequently referenced Egypt in his haute couture collections. His Spring/Summer 2004 collection, titled "Les Marins", included several pieces that directly referenced ancient Egyptian style. The designs featured golden fabrics, pharaonic headpieces, and gowns that echoed the iconic silhouettes of ancient Egyptian rulers. Gaultier's use of metallics and geometric patterns recalled the lavishness of Egyptian royal attire, while the inclusion of headdresses and hieroglyphic motifs underscored his admiration for Egypt's visual language.

Another designer captivated by Egyptian aesthetics was Yves Saint Laurent, whose work often drew from various cultural and historical sources. In 1969, Saint Laurent presented an Egyptian-inspired haute couture collection, which included bold, column-like silhouettes, regal colors such as gold and deep blue, and designs that evoked ancient Egyptian frescoes and statues. His designs channeled the elegance and grace of pharaonic queens, emphasizing long, straight lines and dramatic embellishments, much like the clothing worn by Egyptian nobility.

Chanel's 2019 Metiers d'Art show, under the direction of Karl Lagerfeld, paid tribute to ancient Egypt through extravagant designs that celebrated the civilization's opulence. The collection featured gold lamé gowns, Cleopatra-inspired eyeliner, and accessories modeled after Egyptian collars, hieroglyphs, and scarabs. This show was a perfect example of how modern designers continue to reimagine ancient Egyptian aesthetics within contemporary haute couture.

In addition to haute couture, ancient Egyptian motifs have found their way into more mainstream and streetwear fashion over the decades. Egyptian themes are often reinterpreted to reflect current trends while maintaining a connection to the ancient world. Egyptian-inspired jewelry, particularly those featuring scarabs, ankhs, and hieroglyphs, have had a lasting presence in fashion. The ankh, representing eternal life, is a popular symbol in jewelry, from delicate pendants to large statement pieces. The use of eye of Horus motifs and gold cuffs inspired by pharaonic designs has also remained popular, signifying a blend of spiritual and artistic inspiration.

The singer Rihanna has appeared wearing Egyptian-inspired designs in music videos and on red carpets, popularizing the use of pharaonic themes in contemporary fashion. Brands like Givenchy have also incorporated Egyptian motifs into streetwear lines, including hieroglyphs, eye of Horus symbols, and Cleopatra-inspired makeup looks.

This ongoing fascination with Egypt in fashion can be attributed to the civilization's unique blend of opulence, mystery, and power. Egyptian clothing and accessories were not merely decorative but deeply symbolic, representing spirituality, status, and the connection between humans and the divine. This rich symbolism continues to captivate designers, offering both aesthetic inspiration and a connection to ancient traditions.

4. Art

Ancient Egyptian art has left a significant imprint on modern and contemporary art, influencing some of the most innovative movements of the 20th century. Artists such as Pablo Picasso and Jean-Michel Basquiat found inspiration in the bold forms, distinct symbolism, and stylistic conventions of ancient Egyptian murals and sculptures. Egyptian art, with its unique combination of profile views and frontal torsos, its use of symbolism to convey meaning, and its tendency toward abstraction, resonated with these modern artists, shaping how they approached the human figure, perspective, and narrative in their work.

Though Pablo Picasso is best known for pioneering Cubism, his work exhibits clear ties to ancient Egyptian art, especially in how it handles abstraction and perspective. Picasso's Cubist paintings often depict figures broken down into geometric forms and viewed from multiple angles simultaneously. While ancient Egyptian art didn't display such radical fragmentation, there is a conceptual similarity in the way Egyptian artists represented the human figure by combining multiple perspectives — figures are often shown with their heads in profile, their torsos from the front, and their legs in side view. This stylistic choice in Egyptian art aimed to represent the most recognizable or essential aspects of the human body, much like how Cubism sought to show objects from different perspectives at once.

Picasso's work also reflects Egyptian influences in its symbolic use of imagery. Egyptian art was highly symbolic, with each element carrying layers of meaning — figures, animals, and even colors were tied to mythology and religion. Picasso, while not directly mimicking Egyptian motifs, approached his use of symbols in a similar way. In works like "Guernica" (1937), he employs stylized figures and symbols (such as the bull, horse, and light bulb) to convey deeper emotional and political meanings. This mirrors the use of symbols in Egyptian art, where certain animals or postures had specific connotations related to life, death, and the divine.

Picasso was also influenced by his time in Paris during the early 20th century when there was a significant interest in "Primitivism", a term that European artists used to describe the aesthetics of non-Western cultures, including ancient Egypt and Africa. The fascination with so-called "primitive" forms inspired Picasso's departure from the realism that dominated European art in favor of a more abstract, symbolic, and geometric style — qualities that resonated with the formal structure of Egyptian art.

For Jean-Michel Basquiat, ancient Egyptian art played an essential role in how he expressed ideas related to identity, power, and heritage. A leading figure in the

Neo-Expressionist movement in the late 20th century, Basquiat often drew on African and African Diaspora traditions, with Egyptian culture serving as a key point of reference in his work. Basquiat, who was of Haitian and Puerto Rican descent, sought to highlight African influences in Western art history while also addressing issues of colonialism, racism, and the representation of Black identity.

In several of Basquiat's paintings, he makes direct reference to ancient Egyptian symbols and themes. For example, he frequently used the ankh, the Egyptian symbol for life, as a recurring motif in his work. The ankh served as a powerful statement of cultural pride and a reclaiming of African heritage for Basquiat. This symbol, along with other references to Egypt, allowed him to connect his art to a long tradition of African and African-descended peoples' contributions to world culture.

In his painting "Hollywood Africans" (1983), Basquiat juxtaposes hieroglyphic-like text and symbols with his signature graffiti style to comment on the misrepresentation of African Americans in popular culture and history. The use of hieroglyphs not only references Egypt's revered ancient writing system but also highlights how African cultures have been reduced to symbols in Western art and media. In another work, "Untitled (History of the Black People)" (1983), Basquiat explicitly connects ancient Egypt to African identity, depicting the pharaohs and using African iconography to critique the erasure of Black contributions from history.

Basquiat's layering of these elements reflects the multi-dimensional way ancient Egyptian art often communicated through both image and text. Egyptian murals typically combined hieroglyphs with visual imagery, creating a complex interplay of symbols that served to both narrate and define power dynamics. In Basquiat's work, this layering becomes a way of challenging Western art's historical dismissal of African and Egyptian contributions to civilization.

Ancient Egyptian art's unique approach to representing the human body — particularly its abstracted use of perspective — has had a lasting influence on modernist and contemporary artists beyond just Picasso and Basquiat. Egyptian figures were often depicted in highly stylized, geometric forms, where realism was less important than clarity, symbolism, and function. For instance, the abstract, frontal poses of Egyptian statues and reliefs, with their idealized, simplified bodies, resonate with modern art's emphasis on essential form and minimalism. The iconic seated figure pose seen in statues of pharaohs, where figures appear almost block-like in their rigid, upright posture, echoes the minimalist tendencies of artists like Brancusi or Henry Moore, who sought to reduce the human form to its most fundamental shapes.

The use of scale to denote importance in Egyptian art — a technique where larger figures represent more significant figures or gods — also found parallels in 20th-century works. Artists like Diego Rivera, in his murals, employed a similar hierarchy of scale to depict historical figures in his visual narratives of social and political power. This technique from Egyptian art helped modern artists communicate messages about power, status, and significance.

Beyond individual artists, Egyptian art's influence extends into the broader realms of contemporary visual culture, especially in terms of the symbolic language and the use of iconic Egyptian imagery. For example, the eye of Horus, pyramids, and sphinxes are still regularly referenced in modern design, tattoo culture, music videos, and street art. The persistence of these symbols highlights how deeply embedded Egyptian motifs are in the collective consciousness.

The pyramid, in particular, has become a widely recognized symbol of power, endurance, and mystery. Its abstract, geometric form resonates with modern architecture and design principles, making it a versatile symbol in contemporary art and fashion as well. In modern street art and graffiti culture, Egyptian symbols are often used to convey resistance, protection, or ancient wisdom — much as they did in their original context.

5. Language and Symbolism

Ancient Egyptian symbols have transcended their original cultural context to become iconic in modern times, representing concepts of life, protection, and power. These symbols, deeply embedded in Egyptian mythology and religious practices, have found new life in contemporary culture through jewelry, tattoos, art, and even language.

The ankh is one of the most recognizable symbols from ancient Egypt, and its significance has been remarkably preserved in modern culture. In ancient Egyptian belief, the ankh represented eternal life and was often depicted in the hands of gods or pharaohs in funerary art, symbolizing their dominion over life and death. The shape of the ankh — a looped cross — has been interpreted in various ways, often associated with the union of opposites: male and female, earth and heaven, life and death.

In modern times, the ankh has become a popular symbol in jewelry, tattoos, and fashion, frequently worn as an amulet or a statement piece. Its association with life and immortality has allowed it to cross cultural and religious boundaries, making it a versatile and appealing symbol. Worn by people seeking to express a connection to ancient wisdom, spirituality, or simply as a decorative emblem, the ankh is often seen as a universal symbol of vitality and spiritual endurance.

The ankh's resurgence in the 1960s and 1970s, particularly among members of the counterculture movement and the Black Power movement, further reinforced its use as a signifier of African heritage and identity. For example, leaders like Angela Davis were often seen wearing ankhs, which were used to emphasize a connection to African origins and the deep history of African civilization. This association with cultural pride and the reclamation of African roots has made the ankh not only a decorative symbol but also a political one.

The Eye of Horus (also known as the Wadjet) is another ancient Egyptian symbol that has found a lasting place in modern culture. In Egyptian mythology, the eye represented Horus, the sky god, who lost his eye in a battle with Seth, the god of chaos. The eye was later restored by the god Thoth, making it a symbol of healing, protection, and restoration. In ancient Egypt, it was used as an amulet for protection and to ward off evil.

Today, the Eye of Horus is commonly seen in tattoos, jewelry, and artwork, often adopted for its meaning of protection and well-being. Many people wear or tattoo the Eye of Horus as a protective charm, believed to guard against negative influences, much like its ancient purpose. Its design — a stylized human eye with a distinctive marking — also gives it a unique and enduring aesthetic appeal.

The linguistic influence of ancient Egypt can also be seen in the modern world, with the word "pharaoh" being one such example. The term is derived from the ancient Egyptian title "per-aa" (meaning "great house"), which originally referred to the royal palace itself, but over time came to denote the ruler of Egypt. This shift in meaning likely occurred during the New Kingdom period, as the pharaohs became seen as living gods, and their palace was an extension of their divine authority.

When ancient Egypt became more widely known to the Western world, particularly through interactions with Greek and Hebrew cultures, the term was adopted into these languages. The Bible often refers to the ruler of Egypt as "Pharaoh," and it was later translated into Latin and, eventually, English. Today, the word "pharaoh" is universally understood as the title for the kings of ancient Egypt, even though its origins point to a broader concept related to the physical grandeur of the royal household.

The Sphinx, with its lion body and human head, represents wisdom and strength in ancient Egypt. Today, it is frequently referenced in art, architecture, and literature, often symbolizing mystery or enigma. The Great Sphinx of Giza, in particular, continues to captivate the imagination of people around the world, featuring prominently in discussions about the mysteries of ancient Egypt and its perceived connections to lost knowledge or otherworldly intelligence.

Ancient Egyptian hieroglyphs have inspired modern visual culture, often serving as decorative or symbolic elements in art, tattoos, and design. Hieroglyphs, which were used to record sacred and official texts in ancient Egypt, are still used symbolically to represent ancient knowledge, secrecy, or mysticism in modern contexts. Their geometric simplicity and symbolic meaning have lent themselves to modern graphic design and even logos, where they often appear in stylized forms to evoke the mystique of ancient civilizations.

6. Museums and Global Fascination

Ancient Egypt has secured its place as a cornerstone of global cultural heritage, with its artifacts and art drawing immense public interest. Major museums around the world, such as the British Museum, the Louvre, and the Metropolitan Museum of Art, house extensive collections of Egyptian antiquities, contributing significantly to the phenomenon known as Egyptomania.

The British Museum in London is renowned for its collection of Egyptian artifacts, including the Rosetta Stone and mummies, which have captivated visitors since the museum's establishment in 1753. The museum's Egyptian Gallery houses thousands of artifacts that represent various aspects of ancient Egyptian life, religion, and burial practices. Special exhibitions, such as "Ancient Egypt: The Eternal Experience," have drawn in millions of visitors, showcasing the intricate artistry and technological advancements of civilization.

The Louvre in Paris holds one of the most significant collections of Egyptian antiquities outside Egypt itself, with approximately 50,000 objects spanning several millennia. Its Egyptian Antiquities department features remarkable pieces, including the seated statue of Pharaoh Khafre and the sarcophagus of the priestess of Amun, which provide insights into the religious and social practices of the time. The Louvre frequently organizes special exhibitions that delve deeper into specific aspects of Egyptian culture, such as

"Egypt: Faith After the Pharaohs," exploring the evolution of religious beliefs and practices post-Pharaonic Egypt.

The Metropolitan Museum of Art in New York City is another key institution that has embraced the legacy of ancient Egypt. The museum's Egyptian Art collection is extensive, featuring sculptures, jewelry, and everyday items from various periods of Egyptian history. The Met also offers educational programs and workshops that engage visitors of all ages, helping to demystify the complexities of ancient Egyptian society.

Egyptomania, or the intense fascination with ancient Egyptian culture, has been a recurring theme in Western society, particularly since the 19th century and it extends beyond museum exhibits; it has permeated literature, fashion, art, and film. This cultural phenomenon showcases how ancient Egypt continues to inspire creativity and storytelling in diverse media. Egyptologists like Zahi Hawass now appear in Netflix documentaries and TV programs, making them the modern face of ancient Egypt.

There seems to be no let up in our thirst to learn more about this ancient civilization and its culture, history, and archaeology through their engaging presentations and scholarly insights on various platforms.

"Egyptian mysticism embodies a rich tapestry of esoteric knowledge and spiritual practices that explore the relationship between humanity and the cosmos. The teachings of the ancient Egyptians offer insights into the nature of the soul and the mysteries of life and death."

— James Wasserman

7.1 Egyptian Mysticism and Occultism

Egyptian mysticism and occultism are rich and complex areas that draw upon the beliefs, practices, and symbols of ancient Egyptian religion and philosophy. These traditions have influenced various mystical and esoteric movements throughout history, particularly during the Renaissance and the 19th century when interest in ancient Egypt surged.

Let's explore some of the key concepts, practices, and influences associated with Egyptian mysticism and occultism.

1. The Afterlife and the Soul

The Egyptian view of the afterlife, particularly the journey of the soul through judgment and rebirth, has deeply impacted Western esoteric traditions. The notion that the soul continues to exist beyond physical death resonates with concepts of reincarnation found in several spiritual and mystical philosophies. This idea was revived during the Renaissance, when scholars became fascinated with ancient Egyptian texts and beliefs.

The Hermetic tradition, which emerged in the Renaissance, drew from Egyptian concepts of the soul's journey. Texts attributed to Hermes Trismegistus emphasized the soul's immortality and its quest for enlightenment and unity with the divine, paralleling the Egyptian beliefs in the afterlife and spiritual transformation.

Hermetic practices often involve meditation and contemplation aimed at self-discovery and alignment with divine principles. These practices resonate with the Egyptian emphasis on moral living and spiritual purity, as individuals seek to refine their souls in preparation for the afterlife.

In modern Hermetic rituals, practitioners may invoke Egyptian deities or symbols, such as the ankh or the Eye of Horus, to enhance their spiritual journey and connect with the ancient wisdom embodied in these symbols.

In the late 19th century, the Theosophical Society, founded by Helena Blavatsky, integrated elements of ancient Egyptian beliefs, particularly regarding the afterlife and reincarnation. Theosophists emphasized the cyclical nature of life and the evolution of the soul through multiple lifetimes, mirroring Egyptian beliefs about the journey of the soul after death.

Theosophy also posited that the soul undergoes a journey through various planes of existence, much like the Egyptian belief in navigating the afterlife, where the soul faces trials and judgments before achieving enlightenment.

The rituals of the Theosophical Society often included meditative practices that reflect the journey of the soul and its connection to past lives. By integrating Egyptian symbols and concepts into their practices, Theosophists aimed to draw upon the profound spiritual legacy of ancient Egypt.

The use of talismans and amulets in modern esoteric practices, which have roots in ancient Egyptian beliefs, serves as a means to invoke protection and facilitate spiritual growth, echoing the role of amulets in Egyptian culture to safeguard the soul in the afterlife.

As scholars and practitioners draw from the wellspring of Egyptian wisdom, the quest for understanding the soul's journey remains a central theme in both ancient and modern mystical traditions.

Both Hermeticism and Theosophy demonstrate the significant impact of Egyptian beliefs on spiritual practices, especially concerning personal transformation and moral development.

2. Ma'at – Judgment and Moral Order

The concept of Ma'at, representing truth, justice, and cosmic order, is central to Egyptian beliefs about the afterlife. The weighing of the heart against the Feather of Ma'at not only signifies judgment but also underscores the importance of living a life aligned with truth and morality.

The principles embodied by Ma'at influenced not only the spiritual beliefs of the Egyptians but also had far-reaching effects on subsequent philosophical and religious systems, particularly in their emphasis on ethical living and personal responsibility.

Throughout history, various philosophical traditions have echoed the themes present in Ma'at. The emphasis on truth and justice can be seen in the teachings of philosophers such as Plato, who argued for an ideal society governed by the pursuit of truth and the moral integrity of its rulers. In The Republic, Plato discusses the idea of the philosopher-king, who embodies the virtues of wisdom and justice — qualities that resonate with the values of Ma'at.

The influence of Ma'at also extends into Jewish mysticism, particularly in Kabbalah, which emphasizes the moral and ethical dimensions of existence and the soul's journey. Kabbalistic teachings align closely with the Egyptian notion of living a life that reflects divine principles, where the journey of the soul is intricately linked to the pursuit of justice and truth, mirroring the principles found in Ma'at. Kabbalists believe that the soul must evolve through various stages, engaging in self-reflection and ethical living to align with the divine.

The concept of Tikkun Olam in Kabbalah reflects the idea of "repairing the world" and emphasizes personal responsibility in fostering justice and harmony in society. Much like the Egyptian belief that individuals contribute to cosmic order through their actions, Kabbalistic teachings stress the importance of ethical behavior in achieving spiritual ascension.

In the Kabbalistic model of the Ten Sefirot, we find the divine attributes through which God interacts with the world, encapsulating themes of wisdom, understanding, and justice. Each Sefirah represents a different aspect of divine action, reinforcing the idea that ethical living is essential for spiritual progress. Just as Ma'at governs the order of existence in ancient Egypt, the Sefirot provide a framework for understanding how human actions can align with divine will, emphasizing justice and moral integrity.

The influence of Ma'at persists in modern spiritual movements, where concepts of personal ethics and universal truth continue to be emphasized. This legacy manifests in various contemporary practices that prioritize moral responsibility, self-awareness, and alignment with a greater cosmic order.

Today's ethical movements are increasingly turning to ancient wisdom traditions, such as those exemplified by the concept of Ma'at from ancient Egypt, to inform their perspectives on social justice, environmental stewardship, and personal integrity. This revival highlights a profound recognition that our actions — both individual and collective — are interconnected and have a significant impact on the world around us. Let's explore how these movements draw upon ancient wisdom and the parallels they find in the principles of Ma'at.

- ◆ Social Justice

Social justice movements seek to address systemic inequalities and ensure equitable treatment for all individuals, reflecting the Egyptian emphasis on justice and fairness inherent in Ma'at. The belief that society functions best when everyone is treated with dignity and respect aligns closely with the notion that each person's actions contribute to the greater cosmic order.

Movements such as Black Lives Matter embody these principles by advocating for racial equity, emphasizing the need for justice and accountability in societal structures. This mirrors the Egyptian understanding that societal harmony requires ethical governance and respect for truth.

- ◆ Environmental Stewardship

Environmental movements draw from ancient wisdom to promote ecological sustainability, highlighting the responsibility to care for the Earth. The interconnectedness of life emphasized in Ma'at — where every action affects the cosmic balance — aligns with modern ecological principles.

Organizations such as 350.org and Greenpeace advocate for climate action and sustainable practices,
resonating with the ancient Egyptian belief in the
need to maintain balance and harmony within nature.
This recognition of our interconnectedness fosters a
sense of stewardship and responsibility toward the
environment.

◆ Eco-Spirituality and Ancient Wisdom

Eco-spirituality is a contemporary movement that
merges environmental concerns with spiritual beliefs,
often drawing upon ancient wisdom traditions. This
approach emphasizes the sacredness of the Earth and
the necessity of living in harmony with nature, echoing
the principles of Ma'at. The movement recognizes
that all life forms are interconnected, a theme deeply
rooted in ancient Egyptian thought. The understanding
that one's actions can influence the entire web of life
fosters a sense of responsibility and ethical living.

Many eco-spiritual practitioners invoke practices from
various ancient cultures that emphasize respect for
nature, such as indigenous traditions that honor the
land as sacred. This parallels the Egyptian reverence
for the natural world, where gods and goddesses embodied natural elements and forces.

◆ Balance and Harmony

The concept of balance, central to Ma'at, is reflected
in eco-spiritual practices that advocate for sustainable
living. By seeking a harmonious relationship with the
environment, practitioners aim to restore balance to
ecosystems that have been disrupted by human actions.
Sustainable farming practices, such as permaculture,

are rooted in ancient agricultural wisdom that recognizes the importance of nurturing the land while ensuring its health for future generations. This approach mirrors the Egyptian belief in maintaining harmony with the natural order, where every action is weighed against its consequences.

- ◆ Personal Integrity and Ethical Living

At the core of both Ma'at and modern ethical movements is the emphasis on personal integrity and the moral responsibility of individuals to contribute positively to society. Many contemporary movements encourage individuals to practice mindfulness and make ethical choices that reflect their values. This aligns with the Egyptian principle of living a life of truth and moral integrity, emphasizing that personal actions have broader implications.

The slow food movement, for example, promotes local and sustainable food production, and encourages consumers to be mindful of their choices, fostering a connection between personal health and environmental sustainability. This approach echoes the Egyptian view that personal integrity is vital for the health of the community and the cosmos.

The resurgence of ethical movements drawing from ancient wisdom, particularly the principles embodied in Ma'at, highlights the timeless relevance of these teachings. As contemporary society grapples with complex social and environmental challenges, the emphasis on interconnectedness, balance, and ethical living serves as a guiding principle. By integrating these ancient values into modern contexts, individuals and movements alike are working toward a more just and harmonious world, reflecting the enduring legacy of Ma'at and its profound insights into the human experience.

3. Alchemy and Transformation

The themes of judgment and transformation in ancient Egyptian beliefs significantly influenced the practice of alchemy, intertwining the pursuit of physical transformation with spiritual enlightenment. This relationship is especially evident in how alchemists approached their work, viewing the transmutation of materials as a metaphor for the purification and elevation of the self. Let's explore how these ancient beliefs found fertile ground in the practice of alchemy and the symbolic connections between the two.

At the heart of ancient Egyptian spirituality lies the concept of the soul's journey after death, which is characterized by judgment and transformation. The pivotal moment in this journey occurs during the Weighing of the Heart ceremony, signifying not only judgment regarding one's moral conduct during life but also the potential for transformation into a higher state of existence.

Alchemy, often seen as a precursor to modern chemistry, encompasses both physical and spiritual dimensions. Alchemists sought to transform base metals into noble ones, such as gold, but this endeavor was laden with symbolic meaning.

- ◆ Transmutation as Spiritual Transformation

The act of turning base materials into gold served as a metaphor for achieving spiritual enlightenment and purity. Alchemists believed that, just as physical substances could undergo transformation, so too could the human soul. This transformation involved a process of purification, similar to the soul's journey through judgment in ancient Egyptian beliefs.

The search for the Philosopher's Stone, a legendary substance capable of facilitating this transmutation, parallels the Egyptian belief in the tools (spells and rituals) needed to navigate the afterlife and achieve eternal life. The Stone symbolizes the culmination of spiritual wisdom and transformation.

During the Renaissance, there was a resurgence of interest in ancient wisdom, including Egyptian mythology and alchemical practices. Many alchemical texts from this period referenced Egyptian symbols and concepts, integrating them into the alchemical framework.

- ◆ Symbolism in Alchemical Texts

Renaissance alchemists often drew on the rich symbolism found in Egyptian mythology to convey their teachings. For example, the process of nigredo (blackening), which represents the initial stage of dissolution and purification, echoes the Egyptian concept of confronting and overcoming the darkness of the soul during judgment.

Attributed to Hermes Trismegistus, the Emerald Tablet is a foundational text in alchemy. It encapsulates principles of transformation and contains phrases that resonate with Egyptian beliefs about the unity of the material and spiritual worlds, such as "As above, so below."

- ◆ Alchemy, Spiritual Purification, and Egyptian Journeys

The parallels between alchemical transformation and the Egyptian journey of the soul underscore a shared understanding of personal and spiritual evolution. Just as the deceased must undergo judgment to achieve a blessed afterlife, alchemists viewed their work as a means of refining the self.

The purification process in alchemy often mirrors the Egyptian rituals designed to prepare the soul for the afterlife. Both paths involve a rigorous examination of the self, confronting impurities — whether they be moral failings or base desires — to achieve a state of spiritual enlightenment.

Alchemical rituals, rich with symbolic imagery, often involved the use of sacred geometries, colors, and elements, much like the ritual practices in ancient Egypt, which utilized specific symbols to invoke divine protection and guidance during the afterlife journey.

◆ Legacy of Egyptian Thought in Alchemy

The lasting impact of ancient Egyptian beliefs on the practice of alchemy is evident in how these traditions have interwoven over centuries. Alchemists viewed their quest not merely as a scientific endeavor but as a spiritual pilgrimage, reflecting the transformative power of inner work — a theme deeply rooted in Egyptian thought.

Today, many modern spiritual practices and philosophies, such as New Age movements, continue to draw on the themes of transformation and purification derived from both alchemical and ancient Egyptian teachings. This synthesis illustrates the enduring legacy of Egyptian thought and its profound influence on Western esotericism.

4. Symbolism and Deities

Ancient Egyptian deities and their associated symbolism have had a profound impact on Western mysticism, especially through the revival of interest in esotericism during the Renaissance and in modern occult movements. Many Egyptian gods, concepts, and symbols were adopted, reinterpreted, or syncretized into Western spiritual systems, influencing the development of practices like alchemy, Hermeticism, Theosophy, and other occult traditions. Let's explore how some of the key Egyptian gods played roles in Western mysticism.

- Thoth and Hermeticism

Thoth, the ancient Egyptian god of wisdom, writing, and knowledge, played a significant role in shaping Western esoteric thought, particularly through his association with Hermeticism.

Thoth was later merged with the Greek god Hermes to form Hermes Trismegistus ("Thrice-Great Hermes"). This figure is central to the Hermetic tradition, a body of esoteric philosophical texts and spiritual practices that emerged during the Renaissance. Thoth's attributes of wisdom, magic, and the recording of cosmic laws contributed to the Hermetic emphasis on knowledge, alchemy, and the soul's ascent to divine truth.

One of the foundational Hermetic texts, the Emerald Tablet, attributed to Hermes Trismegistus, contains ideas that echo Egyptian cosmology and metaphysical beliefs. The tablet describes the principle "as above, so below," reflecting the Egyptian idea of cosmic balance, much like the concept of Ma'at, and the alchemical transformation of the self, akin to the spiritual purification seen in the Egyptian afterlife journey.

- Isis in Occultism and Theosophy

Isis, the Egyptian goddess of magic, motherhood, and protection, was widely venerated in the ancient world and became a prominent figure in Western mysticism. As we discovered in the previous chapter, the Cult of Isis spread throughout the Mediterranean world during the Greco-Roman period,- making Isis a key figure in Western esotericism. The Roman mystery religions, which often involved initiation rites and promises of spiritual salvation, were influenced by Egyptian beliefs, particularly the myths surrounding Isis and Osiris, which mirrored ideas of resurrection and immortality.

In the 19th century, Helena Blavatsky, founder of the Theosophical Society, published *"Isis Unveiled"* (1877), a key text in Western occult literature. The title references Isis as a symbol of hidden knowledge, and the book re- flects Blavatsky's integration of Egyptian mysticism into her Theosophical teachings. Theosophy emphasized the idea of spiritual evolution and reincarnation, echoing the themes of death and rebirth found in the mythology of Isis, Osiris, and Horus.

♦ Osiris and the Afterlife in Western Esotericism

Osiris, the god of the afterlife and resurrection, played a crucial role in Egyptian spirituality, which was deeply intertwined with the concept of the soul's immortality. His influence extended into Western mysticism, particularly through ideas related to the afterlife, reincarnation, and spiritual transformation.

The myth of Osiris, where he is murdered, dismembered, and later resurrected, resonated with Masonic ideas about death, rebirth, and the moral transformation of the initiate. Masonic rites often involve symbolism related to Osiris's death and resurrection, emphasizing the soul's journey toward enlight- enment and immortality.

In alchemical traditions, Osiris symbolized the process of spiritual purification and transformation. Alchemists sought to transmute base metals into gold, an effort paralleled by the symbolic transformation of the soul from impurity to perfection. This reflected the myth of Osiris's death and rebirth, which represented the cyclical nature of life and the eventual triumph of spiritual enlightenment over death.

◆ Horus and the Eye of Horus in Mystical Symbolism

Horus, the falcon-headed sky god and symbol of kingship, has permeated Western occult and mystical traditions. The Eye of Horus symbol, representing protection, healing, and restoration, became one of the most recognizable symbols in both ancient and modern esotericism. In Western mysticism, the Eye of Horus has been adopted as a symbol of spiritual insight, protection, and the third eye, which represents inner vision and enlightenment. The symbol is commonly used in occult practices, amulets, and talismans as a means of invoking divine protection and spiritual awareness.

The Hermetic Order of the Golden Dawn incorporated Egyptian symbols and deities into its rituals. Horus, in particular, was viewed as a symbol of the initiate's spiritual awakening and ascent toward divine wisdom. The Eye of Horus was frequently used in their teachings as a symbol of the inner vision that the initiate was meant to develop through esoteric study.

◆ Set and Duality in Western Mysticism

Set, the god of chaos, darkness, and the desert, had a more complex role in Egyptian mythology, often representing disorder and destruction. However, even Set's chaotic nature found resonance in Western esoteric traditions that emphasized duality and the balance of opposing forces.

In Western mystical systems like Hermeticism and Kabbalah, the balance between chaos and order (often symbolized

by Set and Horus in Egyptian myth) became an important theme. Set's role as a necessary counterbalance to Horus's order reflected the idea that chaos was not inherently evil but a part of the natural cosmic cycle. This duality is essential in alchemy, where transformation often involves the reconciliation of opposites.

In the 20th century, The Temple of Set, founded by Michael A. Aquino, explicitly drew upon the figure of Set as a central deity. The temple positioned Set as a symbol of individualism, personal empowerment, and self-deification, diverging from traditional interpretations of Set as a purely chaotic force.

◆   Mystical Practices and Texts

Central to the Book of the Dead are spells designed to protect the deceased from dangers they might face in the underworld, whether from hostile spirits, malevolent forces, or divine judgments. These protective spells reflect a broader magical practice that transcends cultures and time periods, with similar invocations appearing in later Western mystical traditions.

Medieval and Renaissance-era grimoires, or magical handbooks, contain spells, invocations, and rituals often aimed at invoking spirits or divine powers to provide protection, guidance, and success in the practitioner's endeavors. Many of these grimoires, such as the "Key of Solomon" or "The Lesser Key of Solomon" (17th century), share conceptual similarities with the spells of the Book of the Dead, particularly in their focus on ensuring safe passage through spiritual trials and invoking higher powers for assistance. The Egyptian notion of using magical words to secure the soul's safety in the afterlife resonates with these later traditions.

The Book of the Dead was not only about protecting the deceased but also about guiding the soul through the complex geography of the afterlife. The idea that the soul embarks on a journey, encountering trials and judgments, parallels later esoteric traditions that emphasize the transformative journey of the soul toward enlightenment or spiritual fulfillment.

Many Western mystical systems, including Kabbalah, Hermeticism, and Theosophy, adopted the notion of the soul's journey through different spiritual realms. These traditions often map out stages of spiritual transformation, resembling the Egyptian concept of the soul's progression through the underworld. In Kabbalah, for instance, the soul ascends through the Sephirot (ten emanations of the divine), a spiritual ladder toward unity with God. Similarly, in Hermeticism, the soul's ascent through celestial spheres is key to its transformation and reunification with the divine source. Both reflect the Egyptian model of the soul overcoming challenges and judgments in the afterlife, ultimately achieving immortality.

Today, Egyptian mysticism has found a place in various New Age and occult practices. Many modern practitioners draw on ancient Egyptian symbols, rituals, and deities to develop spiritual practices, emphasizing personal transformation and empowerment. The Hermetic Order of the Golden Dawn, a late 19th-century magical order, incorporated Egyptian symbols and rituals into its teachings, emphasizing the importance of Egyptian mythology in the practice of magic.

Some contemporary spiritual movements seek to reconstruct ancient Egyptian religious practices, blending them with modern beliefs and practices. These reconstructionist traditions emphasize the worship of Egyptian deities, rituals, and the importance of community. The Kemetic Orthodox community, for example, practices a modern form of ancient Egyptian religion, focusing on the worship of the gods and goddesses of ancient Egypt, ritual practices, and the study of ancient texts.

The fascination with Egyptian mysticism and occultism has permeated literature, art, and popular culture. Writers and artists have drawn upon Egyptian themes to explore concepts of magic, mystery, and the unknown.

From ancient religious beliefs centered on the afterlife and deities to the influence of Egyptian thought on modern esoteric traditions, the impact of ancient Egypt on spirituality and mysticism is profound. This rich legacy continues to inspire and shape contemporary spiritual movements, ensuring that the allure of ancient Egyptian wisdom remains vibrant in the modern world.

5. Egyptian Mythology and Modern Psychological Archetypes

Carl Jung's theory of archetypes and the collective unconscious is another example of a nuanced modern influence stemming from ancient Egypt. While Jung did not explicitly cite Egyptian mythology as his primary source, his idea that certain symbols recur across cultures aligns with how Egyptian gods and symbols — like Isis as the archetypal mother or Osiris as the archetype of death and resurrection — have continuously appeared in Western culture.

The narrative structure of the Osiris myth, where Osiris is killed, dismembered, and resurrected by his wife Isis, closely mirrors the hero's journey archetype, later formalized by Joseph Campbell. This myth has subtly influenced modern storytelling frameworks, from literature to cinema, where themes of death, transformation, and rebirth (whether literal or metaphorical) echo this ancient Egyptian story.

6. Egypt and Afrocentrism

In the 20th century, Afrocentric thought reasserted the importance of ancient Egypt as a foundational African civilization. Intellectuals like Cheikh Anta Diop argued that Egypt was fundamentally African, and that its cultural and scientific achievements were part of a broader African legacy. This challenged the traditional Eurocentric historical narratives that had often downplayed Egypt's African roots.

Afrocentric scholars and movements viewed Egypt as the cradle of African culture, knowledge, and innovation, seeking to reclaim it from centuries of colonial historiography that had obscured its African identity. This has had significant cultural and political ramifications, influencing African and African American identity, pride, and movements like Pan-Africanism.

The nuanced influences of ancient Egypt stretch far beyond its visible impact on architecture, fashion, and film. Historically, Egyptian thought has permeated intellectual movements such as psychoanalysis, Renaissance humanism, and revolutionary ideals, while its symbols and mythology have shaped modern spiritual practices and psychological theories. This deeper cultural and philosophical legacy continues to influence how modern societies perceive ancient wisdom and its relevance to contemporary issues.

"The construction of the temples and monuments of ancient Egypt showcases a remarkable blend of artistry and engineering. The precision with which they were built continues to astonish modern architects and archaeologists alike."

— Aidan Dodson

7.2 Engineering Marvels: Lessons from the Past

Visitors to Egypt are usually struck by the sheer scale of the ancient structures and their precision, at a loss to understand how such accomplishments were even possible four or five thousand years ago.

It's important to mention here that the theory that the pyramids were built by slaves has largely been debunked by modern archaeology and historical research. The popular notion, often perpetuated by early historians and Hollywood films, that the pyramids were constructed by an enslaved workforce is no longer widely accepted by scholars. Instead, evidence shows that the workers who built the pyramids, particularly the Great Pyramid of Giza, were skilled laborers who were well-fed, housed, and organized into a sophisticated workforce.

Archaeological excavations near the Giza pyramids have uncovered the tombs of pyramid builders in the 1990s that provide compelling evidence the laborers were not slaves but respected workers. The fact that these individuals were buried in close proximity to the pyramids, in well-built tombs, suggests they were held in high regard. Slaves, particularly in ancient societies, would not have been granted such honors.

From what we know now, Egyptian society had a system where able-bodied men were required to work for the state as part of a corvée system. This was a form of tax payment through labor, where peasants worked on large state projects like the pyramids for a portion of the year, especially during the flood season when farming was not possible. The workers would rotate in and out, meaning they were not enslaved but conscripted laborers performing civic duty for the state.

Archaeologists have also found worker villages that show evidence of well-constructed living quarters, bakeries, breweries, and even medical care. This suggests that the workforce was cared for and nourished. Workers were given good-quality food, including meat, and had access to medical treatment, which would have been unlikely for enslaved people. These settlements are indicative of a labor force that was important to the state and their welfare prioritized to ensure the success of the massive projects.

In fact, it was essential to have skilled laborers, such as engineers, architects, stone masons, and other specialists for the construction of the pyramids and temples. Teams of such workers competed to complete their sections of the pyramid, and inscriptions found at the site indicate that workers took pride in their contributions, again pointing to a more voluntary and respectful form of labor than slavery.

From the construction of the pyramids to their irrigation systems and monumental temples, Egypt's advancements reveal a mastery of materials, mathematics, and labor organization that modern societies still seek to emulate. Below we will take a look at the key engineering lessons derived from ancient Egypt and their relevance today.

1. Precision in Large-Scale Construction

The most iconic of Egypt's engineering marvels is undoubtedly the Great Pyramid of Giza, built during the reign of Pharaoh Khufu (around 2580–2560 BCE). This ancient structure, originally standing at 146.6 meters (481 feet), remained the tallest man-made structure in the world for nearly 4,000 years. Despite its immense size, the pyramid was constructed with extraordinary precision.

- ◆ The layout of the Great Pyramid's base is almost perfectly aligned with the cardinal directions (north, south, east, and west). Modern engineers remain in awe of this feat, achieved without modern tools. The pyramid's proportions also demonstrate the Egyptians' understanding of geometric principles that are now foundational in modern engineering.

- ◆ The limestone blocks used in construction, some weighing as much as 80 tons, were transported from quarries miles away. The sheer scale of the operation highlights the Egyptians' ability to coordinate large workforces and resources. Modern lessons can be drawn from their methods of material transport, which involved a combination of sledges, ramps, and water to ease friction — a technique that continues to be studied by modern civil engineers exploring innovative methods for material transportation in large projects.

2. Sustainable Irrigation Systems

One of the greatest legacies of ancient Egyptian engineering is their sophisticated system of irrigation. The annual flooding of the Nile River allowed for the creation of rich agricultural lands, but the Egyptians took advantage of this natural phenomenon by building canals, basins, and reservoirs to control and optimize water distribution for agriculture.

- The construction of nilometers (structures used to measure the water level of the Nile) allowed the Egyptians to predict flooding and plan their agricultural cycles accordingly. The use of channels and basins for water storage in periods of low inundation has parallels to modern-day practices of water conservation and sustainable irrigation. Today's engineers are studying ancient Egyptian methods as examples of early water management systems, particularly in areas prone to drought or water scarcity.

- In an era where water scarcity is a global concern, Egypt's ancient irrigation practices offer insights into sustainable water use. Countries like Egypt and others in arid regions are looking to historical methods to design modern irrigation systems that conserve water while ensuring food security.

3. Monumental Architecture and Urban Planning

Beyond the pyramids, Egypt is famous for its temples, such as Karnak and Luxor, which showcase complex engineering and architectural expertise. These structures were designed to withstand the test of time, using durable materials like sandstone and granite. The alignment of these temples with celestial events, such as the solstices, also reflects a profound understanding of astronomy and its integration into architectural design.

- The construction of columns, walls, and obelisks, often weighing hundreds of tons, required sophisticated tools and techniques. The precise fitting of enormous stone blocks without mortar remains a marvel. Modern engineers can draw lessons from the Egyptians' understanding of weight distribution, friction, and leverage — principles still vital in contemporary architecture and large-scale construction.

◆ Egypt's planned cities, like Amarna, were designed with specific spatial relationships, integrating temples, palaces, and residential areas. Modern urban planning draws inspiration from such ancient layouts, particularly in terms of the integration of public spaces with natural landscapes and the balance between religious, administrative, and residential functions in a city.

4. Solar Orientation and Green Energy Potential

The orientation of many of Egypt's temples and pyramids is aligned with solar events. For instance, the axis of Karnak Temple was aligned to the summer solstice sunrise, allowing sunlight to penetrate deep into the temple during certain times of the year. Similarly, the Abu Simbel temple was built so that twice a year, the sun's rays would illuminate the statues inside the inner sanctuary.

◆ This careful orientation to celestial phenomena reflects a deep understanding of the environment. Today, solar orientation is a key consideration in sustainable building practices. Architects and engineers design buildings to maximize natural light and reduce energy consumption, drawing parallels to the ancient Egyptians' alignment of structures with the sun for religious, agricultural, and functional purposes.

◆ The study of ancient Egypt's solar-based alignments has also inspired modern applications in solar energy. By examining how ancient buildings were oriented to make use of natural light and heat, modern architects are integrating passive solar designs into contemporary structures, which can reduce energy usage and improve sustainability.

5. Scalability and Labor Organization

The sheer scale of Egyptian construction projects, especially the pyramids, necessitated a highly organized workforce. Contrary to early misconceptions, Egypt's large construction projects were not built by slaves, but by a well-fed and organized labor force. Work gangs, each with specialized roles, were responsible for various stages of construction, from quarrying stones to transporting them and placing them on site.

- ◆ The Egyptians created a system of rotating labor, where workers would spend part of the year contributing to large construction projects and the rest of the year tending to their farms. This system of organized, skilled labor resembles modern project management practices, where large-scale projects are broken down into stages and tasks are distributed according to skills and availability.

- ◆ Modern construction and project management have much to learn from the Egyptians in terms of logistics and labor organization. Contemporary lessons on productivity, worker welfare, and task specialization can be derived from these ancient practices. Large infrastructure projects today — such as the construction of skyscrapers, dams, or even space exploration — require similar coordination of vast resources, specialized skills, and efficient management.

Today's engineers and architects are increasingly revisiting these ancient methods, drawing inspiration from their environmental awareness, material mastery, and ability to solve complex problems without the benefit of modern technology. The legacy of Egypt's engineering marvels remains not only in the monuments themselves but in the enduring principles of design, sustainability, and human organization that continue to inform modern engineering practice.

6. Ancient tools and modern uses

We also have the ancient Egyptians to thank for laying the groundwork for a variety of tools and technologies still utilized today. Here's a quick look at some of them:

- Merkhet: One of the earliest known astronomical instruments, the merkhet, was used by the Egyptians to track the movements of stars. This tool helped align temples and monuments with celestial bodies. It functioned as a sort of plumb line and allowed the Egyptians to establish precise alignments in their constructions based on star observations.

- Clepsydra (Water Clock): The Egyptians also developed the water clock, a device that used the flow of water to measure time, particularly at night. This innovation allowed them to measure the passage of hours without relying on sunlight or celestial bodies, contributing to their timekeeping and daily rhythms.

- The Plumb Bob: A simple yet essential tool consisting of a weight attached to a string, used to determine vertical alignment. In ancient Egypt, plumb bobs were instrumental in the construction of monumental structures like pyramids and temples, ensuring that their walls were straight and vertical.

Today, the plumb bob remains a fundamental tool in construction and carpentry, helping to ensure precision in building and structural design.

- The Lever: A basic machine that amplifies force, enabling a heavy load to be moved with less effort. Ancient Egyptians employed levers to transport large stone blocks during the construction of their pyramids and temples, playing a crucial role in lifting and positioning heavy materials. The principle of the lever is foundational in physics and engineering, and modern cranes and hoists utilize these principles to lift heavy objects efficiently.

- The Shadoof (Shaduf): A hand-operated device used for lifting water, characterized by a long pole with a bucket on one end and a counterweight on the other. Egyptians utilized shadoofs to irrigate their fields, drawing water from the Nile, particularly during the inundation period. Variations of the shadoof are still employed in agricultural practices worldwide today, with the concept of using a counterweight to facilitate lifting evident in modern irrigation systems and pumps.

- The Sickle: A curved blade designed for harvesting crops. The ancient Egyptians used sickles to efficiently cut grains like wheat and barley, which were staples in their diet. This tool has evolved over time into modern harvesting tools like the scythe and mechanical harvesters, which are crucial for contemporary agriculture and efficient grain harvesting.

- Mortar and Pestle: A tool utilized for grinding and mixing various substances, including food and medicinal ingredients. Ancient Egyptians used these tools to grind grains, herbs, and minerals for both cooking and medicinal purposes. Today, the mortar and pestle remain common kitchen instruments, used in cooking and pharmaceuticals for grinding spices and mixing ingredients.

◆ Papyrus: A plant-based material that served as a writing medium in ancient Egypt. The Egyptians developed papyrus to keep records, write literature, and document religious texts. It is considered a precursor to modern paper, with the techniques for making and using it evolving into the paper we utilize today.

◆ Surgical Instruments: A variety of surgical tools, including scalpels, forceps, and probes. Egyptian physicians demonstrated a significant understanding of anatomy and medicinal practices through various surgical procedures. Many surgical instruments in modern medicine trace their designs back to these ancient tools, with the principles of hygiene and surgical practice developed in ancient Egypt forming the groundwork for contemporary medical practices.

◆ The Ramp: A sloped surface that connects two different elevations. The ancient Egyptians used ramps to transport large stone blocks to higher levels during the construction of their pyramids. The concept of ramps remains integral to modern construction and architecture, as they are utilized for accessibility (such as ADA compliance), material movement, and play a crucial role in engineering and design.

- ◆ Cosmetic Containers and Mirrors: A variety of containers for cosmetics and utilized polished metal or stone as mirrors. Cosmetics held significant importance in Egyptian culture for both beauty and religious purposes. The design of cosmetic containers and the use of reflective surfaces in cosmetics have evolved, but their roots can be traced back to these ancient practices, influencing modern packaging and beauty standards.

The innovations of ancient Egypt not only reflect their ingenuity and understanding of their environment but also laid essential groundwork for various technologies we utilize today. By observing the natural world and creating tools that addressed their needs, the Egyptians made contributions that resonate through history, influencing contemporary practices across various fields, including construction, agriculture, timekeeping, and medicine.

In the grand mosaic of human history, the vibrant pieces of ancient Egypt continue to shine brightly in the modern world, reminding us of a civilization that reached for the stars while remaining deeply connected to the earth. From the majestic pyramids that still astound us with their architectural genius to the profound wisdom embedded in their mystical texts, the echoes of this remarkable society resonate through our tools, technologies, and spiritual philosophies.

Ancient Egyptian innovations in engineering, agriculture, and medicine laid the groundwork for countless advancements, while their rich symbolism and mythology inspire contemporary art, fashion, and ethics.

I began this chapter with a question and I think we are near to finding our answer:

As we stand in awe of the legacy of ancient Egypt, we are invited to explore the depths of our own existence, embracing the timeless wisdom of a culture that sought to understand life, death, and the cosmos — a civilization that, even millennia later, ignites our imagination and beckons us to uncover the mysteries of our own journey!

Moving forward

Reflecting on everything I've learned from Egypt's ancient civilization, I am deeply moved by the scale of its accomplishments — its architectural wonders, spiritual depth, and its ability to leave an indelible mark on human history. The grandeur of the pyramids, the intricate carvings on temple walls, and the sophistication of hieroglyphs serve not merely as remnants of a bygone era but as lasting symbols of what humanity can achieve when driven by a profound blend of ingenuity, faith, and collective purpose. Egypt's ancient society mastered the art of immortalizing its culture, beliefs, and wisdom, leaving behind a legacy that continues to inspire people thousands of years later. They understood not only the material world but also the mysteries of life, death, and the cosmos, and they sought to capture these insights in ways that transcended time. The enduring relevance of Egyptian achievements reminds us of the far-reaching potential of human civilization to create, understand, and shape the world.

What strikes me most, however, is the cyclical nature of history. Civilizations rise, evolve, and sometimes fall, but often they find new ways to emerge again, drawing on the strengths of their past while adapting to the present. My hope for Egypt is that this cycle of innovation and greatness will repeat itself. The spirit of creativity, wisdom, and resilience that once defined this land can return, this time perhaps in new forms — whether in science, technology, culture, or social advancement. Egypt's ancient achievements need not be seen as a distant past but rather as part of a larger, recurring cycle of human progress. With the right vision, the blend of historical richness and forward-thinking innovation could lead Egypt to new heights. I believe the greatness that once flourished along the banks of the Nile is not lost, but dormant, waiting to reawaken. Egypt can once again be a center of global influence, drawing upon its profound past to inspire new generations, contributing to a world where ancient wisdom meets modern progress.

As history flows like the Nile, civilizations rise with new strength from ancient roots. Egypt's past is not just a memory but a seed for the greatness yet to come.

"In the silence of the heart, where chaos gives way to clarity, the essence of Ma'at reveals itself—not merely as order and truth, but as the cosmic dance of energies harmonizing the soul with the universe."

— Dr. Leo Rastogi

CONCLUSION

In ancient civilizations, from the Egyptians to the shamans of Africa and the Americas, leaders were revered as spiritual intermediaries with profound connections to the divine. These figures were not merely rulers or healers but conduits between the human and cosmic realms, guiding their communities with both temporal power and spiritual wisdom. Egypt, in particular, stood as a beacon of civilization, advancing architecture, mathematics, medicine, and astronomy. Yet, perhaps its greatest legacy was its spiritual framework, where the Pharaohs embodied a deep connection to the divine, blending material progress with spiritual depth.

Today, Egypt's ancient wisdom still echoes through time, yet the world it helped shape seems to have lost touch with that spiritual foundation. In our modern culture, those once seen as divine leaders have been replaced by tech giants, influencers, and celebrities. These contemporary figures dominate our daily lives, not by channeling ancient wisdom or offering spiritual guidance, but by wielding control over digital platforms and consumer attention. We now look to technology for solutions to life's mysteries and challenges, yet it feels as though something essential has been lost in this shift.

In the Middle East, a region once rich in sacred knowledge, religious wars rage, driven more by rigid dogma and political agendas than by true spiritual insight. The harmony and balance — the Ma'at — that Pharaohs once sought have been replaced by the pursuit of domination and control. The region that used to be at the heart of the world's spiritual evolution, is now mired in violence, often justified in the name of faith.

As we watch these conflicts unfold, it is worth reflecting on how disconnected we have become from the ancient wisdom that prioritized unity, balance, and the greater good. The spiritual leaders of Egypt sought divine guidance to navigate chaos, yet today's conflicts are often the result of dogmatic interpretations imposed on complex human issues. Egypt's legacy reminds us that true progress and civilization come from a balance of the earthly and the divine — a balance sorely needed in our fractured world.

This disconnection extends beyond global conflicts to our personal lives. In a culture obsessed with the pursuit of happiness, we often treat it as something to be attained at all costs — measured by achievements, wealth, or status. In our relentless chase for success, we lose sight of the inner harmony that once grounded civilizations.

The ancient Egyptians understood that true contentment could not be found in material gain alone. They sought wisdom that transcended the physical world, understanding that happiness is not a prize to be won but a state of being to be nurtured. Their pursuit of balance — both personally and societally — offers us a model for navigating our own tumultuous times.

There has never been a better time than now to revisit the ancient paths trodden by the mystics and seers of old. They recognized that life was more than what meets the eye, that our existence is connected to forces greater than we can comprehend. By turning inward, embracing self-reflection, and reconnecting with the deeper wisdom of the universe, we can find clarity in a world that feels increasingly divided.

The wisdom of Egypt offers a profound lesson: progress and happiness arise when the material and the spiritual are not separate but intertwined. It is a balance we must reclaim if we are to navigate the challenges ahead, both as individuals and as a society.

As you reflect on the ancient wisdom of Egypt and the timeless lessons it offers, may you find inspiration to seek balance in your own life. The path of spiritual discovery is unique to each of us, but it is a journey worth taking. Whether you look inward for answers or seek guidance from the universe, may you walk forward with an open heart, embracing both the material and the divine. I wish you well on your spiritual journey, trusting that you will uncover the harmony and deeper truths that lead to a life of fulfillment and peace.

The mystique of ancient Egypt endures, its enigmatic connection to the spiritual realm offering a glimpse into a world where the boundaries between the earthly and the divine are seamlessly intertwined.

About the author

Dr. Leo Rastogi is a student of life, whose journey seamlessly blends entrepreneurship, philosophy, leadership, research and spirituality. His diverse education, profound spiritual experiences since childhood, and global exploration have shaped his unique world view and insights.

Having traveled to 70+ countries, Leo has delved into ancient traditions, esoteric sciences, quantum physics, and abstract mathematics. He is a distinguished alumni of Harvard Business School, King's College, Oxford, and Salford University, and holds a PhD in Leadership. Leo has also studied psychology, computer science, and neuroscience

As a respected authority on Conscious Leadership, Leo's immersion in monastic practices and his encounters with enlightened masters have given him an especially rich perspective. He bridges the realms of science and spirituality, inspiring scholars, seekers, and students alike. Apart from being an author and much-sought-after public speaker, he is a serial technology and wellness entrepreneur who passionately explores the convergence of emerging tech, science, spirituality, ancient wisdom, and global well-being.

Leo has also served on boards of global business and won numerous awards, as well as contributed to leading publications. He has synthesized much of this wisdom into his critically acclaimed books: "**Many Paths, Many Truths: A Journey through World Religions that have Shaped 20th Century Humanity**" and "**To Discover or to**

Believe: A Journey through what Science does not answer and Spirituality does not ask." He is also the author of "**Ayam: Discover your Beautiful Self**" and "**Life is Good: Conversations about Joy, Purpose, and Fulfillment**".

Leo's life mission is to contribute to transforming human well-being with a synthesis of cutting-edge science, ancient wisdom, and a profound understanding of the human experience.

Appendix: Resources for Further Exploration

Adams, M., *Meet Me in Atlantis: My Quest to Find the Legend of Atlantis,* Dutton, 2015.

Alcott, L. M., *Lost in a Pyramid, or The Mummy's Curse*, 1869.

Allen, J. P., *Middle Egyptian: An Introduction to the Language and Culture of Hieroglyphs,* Cambridge University Press, 2010.

Bauval, R., and A. Gilbert, *The Orion Mystery: Unlocking the Secrets of the Pyramids,* Crown Publishing Group, 1994.

Bauval, R., *The Egypt Code,* Century, 2006.

Baines, J., and J. Malek, *Atlas of Ancient Egypt, Facts on File*, Time Life UK, 1980.

Blavatsky, H. P., *The Secret Doctrine*, CreateSpace, 2011

Brennan, H., Atlantis: *The Legend of a Lost Civilization*, The Book People, 1999.

Churton, T., *Gnosticism: New Light on the Ancient Tradition of Inner Knowing*, Inner Traditions, 2002.

Cayce, E., *Edgar Cayce on Atlantis*, A.R.E. Press, 1994.

Clarke, J. H., *African People in World History*, A & B Books, 1993.

Cotterell, M., *The Tutankhamun Prophecies: The Sacred Secret of the Maya, Egyptians, and Freemasons*, Bear & Company, 1999.

David, R., *Handbook to Life in Ancient Egypt*, Facts on File, 1998.

DeConick, A., *The Gnostic New Age: How a Forgotten Spiritual Tradition Can Change Your Life*, Continuum, 2007.

Dickinson, O., *The Aegean Bronze Age*, Cambridge University Press, 1994.

Donnelly, I., *Atlantis: The Antediluvian World*, Harper & Brothers, 1882.

Doyle, A. C., *Elementals and the Curse of Tutankhamun*, 1922.

Eliade, M., *The Myth of the Eternal Return: Cosmos and History*, Princeton University Press, 1954.

El-Tawil, Sherif et al., *Lord Carnarvon's death: the curse of aspergillosis?* The Lancet, Volume 362, Issue 9386, 836.

Emmel, S., *The Nag Hammadi Library: A Historical and Literary Introduction*, E.J. Brill, 1999.

Fagan, B. M., *The Long Shadow of the Roman Empire: Cultural Adaptation in the Western Empire*, Cambridge University Press, 2013.

Finkelberg, M., *The Minoan World*, Cambridge University Press, 2015.

Frankl, V. E., *Man's Search for Meaning*, Beacon Press, 2006.

Friedan, B., *The Secret Life of the Ancient Egyptians: A Mystical Journey*, Shambhala Publications, 2004.

Haigh, M., *The Mysteries of Ancient Egypt: How They Still Influence Our World Today*, North Atlantic Books, 2004.

Hancock, G., *Fingerprints of the Gods: The Evidence of Earth's Lost Civilization*, Crown Publishing Group, 1995.

Hancock, G., *The Sign and the Seal: The Quest for the Lost Ark of the Covenant*, Touchstone, 1992

Hassan, F. A., The Impact of the Environment on the Social Structure of Ancient Egypt, University of California Press, 1997.

Hawass, Z., *Experiences with Tomb Curses During Excavation,* 2021.

Holland, S., *The Gods of Ancient Egypt*, Amberley Publishing, 2014.

Hornung, E., *The Secret Lore of Egypt: Its Impact on the West,* Inner Traditions, 1999.

Kemet, R. U. N., *The Metu Neter: The Great Oracle of Tehuti and the Egyptian System of Spiritual Cultivation*, Tehuti Publishing, 1999.

King, K. L., *The Gospel of Mary of Magdala: Jesus and the First Woman Apostle*, Polebridge Press, 2003.

Kramer, S. N., *History Begins at Sumer: Thirty-Nine Firsts in Recorded History*, University of Pennsylvania Press, 1981.

Lehner, M., *The Complete Pyramids: Solving the Ancient Mysteries*, Thames & Hudson, 1997.

Loudon, J. C., *The Mummy!: Or a Tale of the Twenty-Second Century*, H. Colburn, 1827.

Marinatos, S., *Thera and the Aegean World, The Thera Foundation*, 1987. (Originally published in Greek)

Meyer, M., *The Gnostic Gospels of Jesus: The Mystical Teachings of the Nag Hammadi Library*, Inner Traditions, 2005.

Milton, G., *The Gods of Eden*, 1993.

Naunton, C., *Egyptian Myth: A Very Short Introduction*, Oxford University Press, 2016.

Naydler, J., *Shamanic Wisdom in the Pyramid Texts: The Mystical Theology of Ancient Egypt,* Inner Traditions, 2005.

Pagels, E., *The Gnostic Gospels,* Vintage Books, 1989.

Plato, *Timaeus and Critias*, Harvard University Press, 1929.

Redford, D. B., *The Oxford History of Ancient Egypt,* Oxford University Press, 2001.

Riley, G. J., *One Jesus, Many Christs: How Jesus Inspired Diverse Expressions of Faith*, HarperOne, 1997.

Ronn, E., *Atlantis: The Evidence of the Ancient Greeks*, Hill and Wang, 2019.

Robinson, J. M., *The Nag Hammadi Scriptures: The Revised and Updated Translation of the Gnostic Scriptures*, HarperOne, 1990.

Sitchin, Z., *The 12th Planet: Book I of the Earth Chronicles*, Harper & Row, 1976.

Todorov, T., *The Conquest of America: The Question of the Other*, Oxford University Press, 1999.

Van Dijk, J. T., *Ancient Egyptian Literature: An Anthology*, Brill, 2000.

Vassiliki, Tzanavari, *Santorini: A Brief History,* Vassiliki Tzanavari Publishing, 2015.

Wilkinson, T., *The Rise and Fall of Ancient Egypt*, Random House, 2010.

Wilkinson, R. H., *The Complete Gods and Goddesses of Ancient Egypt,* Thames & Hudson, 2003.

Wood, M., *In Search of the Trojan War,* University of California Press, 1996.

All images sourced on Shutterstock.

Other titles
by Dr. Leo Rastogi

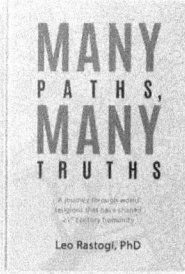

Many Paths, Many Truths:
A Journey through World Religions that
have Shaped 20th Century Humanity

Ayam: Discover your
Beautiful Self

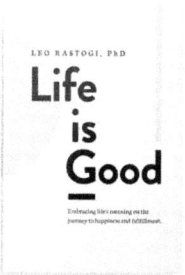

Life is Good: Conversations about Joy,
Purpose, and Fulfillment.

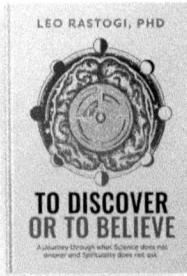

To Discover or to Believe:
A Journey through what Science
does not answer and Spirituality
does not ask

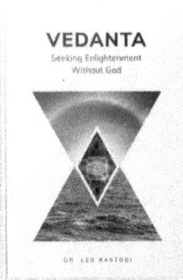

Vedanta: Seeking Enlightenment
Without God